I0045019

T. Lloyd

Bimetallism Examined

T. Lloyd

Bimetallism Examined

ISBN/EAN: 9783743384071

Manufactured in Europe, USA, Canada, Australia, Japa

Cover: Foto ©berggeist007 / pixelio.de

Manufactured and distributed by brebook publishing software
(www.brebook.com)

T. Lloyd

Bimetallism Examined

Bimetallism Examined.

By T: LLOYD.

REPRINTED FROM THE "Statist."

"STATIST" Press, 51, Cannon Street, London.

1894.

WYMAN AND SONS, LIMITED,
PRINTERS,
63, 65, & 67, CARTER LANE, DOCTORS COMMONS,
LONDON, E.C.

PREFACE.

THE following papers have all appeared in *The Statist*, and are now collected and republished, in the hope that they may be of service to those who, without being professed students of currency questions, yet desire to understand the Bimetallist controversy. The method of discussion, I am painfully aware, has grave drawbacks. It makes systematic treatment impossible. It imposes such limits as render exhaustive analysis and exposition unattainable. And it demands brevity so strongly that it tempts to an omission of necessary qualifications. But I venture to think that it has counterbalancing advantages. It compels the writer to avoid technical language, which too often only serves as a veil to hide poverty of thought. It requires him to use plain words intelligible to everybody. It obliges him to avoid prolixity, to disregard what is unessential. It urges him to array the strongest and most telling arguments; and it does not impose such a strain upon the attention of the reader as a book does. The articles having originally appeared at intervals of a week apart, and having been addressed to a more or less varying body of readers, had to be complete, each in itself; or, at all events, had to be intelligible without reference to what had gone before and therefore needed to have a beginning, middle and end. Consequently the reader may take up or lay down the pamphlet without losing the thread of the argument. Yet I venture to hope that there will be found enough of continuity of thought running through the papers to justify their republication in the present form,

T. Ll,

BIMETALLISM.

REPRINTED FROM "THE STATIST."

THE THEORY OF VALUE.*—I.

THE accepted theory of value, I venture to think, is largely responsible for the discredit into which political economy has fallen; it confuses thinkers and it misleads the public. How widespread is the dissatisfaction with it is proved among other things by the keen attention excited by the speculations of the Austrian school. The theory briefly stated is that value depends upon supply and demand conditioned by the cost of production. The cost of production, however, does not directly affect value. It determines whether a thing is or is not to be produced in the future, but, except very indirectly, it has no bearing upon value. If the cost of production is increased, either a substitute will be found and the thing will go out of use, or production will be transferred to some other place where the cost is less—provided the desire for it is not keen. But if there is a very strong desire for it the value, of course, will increase. In that case, however, the value is determined by the strength of the desire, not by the cost of production. In short, the introduction of the cost of production is supereroga-

---

* From *The Statist* of November 11, 1893.

B

tory if the rest of the theory is correct. The first objection to the theory that will naturally occur is that supply is a physical quantity and demand a mental state, and that there cannot be a quantitative relation between a physical quantity and a mental state. A non-natural meaning, therefore, must be given to the word "demand" to enable the theory to bear criticism. But even if this be done the theory is unsatisfactory, for every day's experience proves that it does not fit in with the facts. It is not, for example, the supply actually offering in the market, nor even the supply known to exist and believed to be saleable, that alone affects value. The prospect of an increase or a decrease in the supply often has a more powerful influence, as, for example, the state of the growing crops. To make the theory, therefore, even approximately accurate, instead of "supply" simply we should say the supply actual, potential, and prospective. But the potential and prospective supply at all events can only be estimated, and, therefore, it is not supply, but the estimate of supply, that affects value. Now let us turn to the word "demand," and the first thing that strikes us is that the very strongest of all demands has no influence whatever upon value—as, for instance, the demand of the penniless all over the world for food, clothing, and dwellings. To get over this objection it has been suggested that for the word "demand" should be substituted the phrase "effective demand," which is very much like saying that a demand affects value *when* it affects value. What is, of course, meant by the phrase is the purchasing power of those who demand. But the purchasing power can be nothing more than the supply of other things which those who demand have at their disposal, so that our analysis of the theory brings us to this, that the value of a thing is determined by the relation between the estimate of the supply of that thing and the estimate of the supply of all other things offering for it. Can there be a quantitative relation between two guesses?

It may be objected by those who hate to be disturbed in their opinions that the above is mere hair-splitting. I submit that it is much more, and I proceed to show that the accepted theory does not account for the facts. Between 1865, when the Civil War ended, and 1879, when resumption was effected, there was a marvellous growth of wealth and population in the United States. On the other hand, from 1870 downwards, there was a very considerable reduction in the incontrovertible paper money, yet that money did not rise to its legal par until the very eve of resumption, the very end of 1878. Then the success of the measures taken for carrying out resumption, and of the negotiations for the re-funding of the interest-bearing debt so powerfully impressed the imagination of the world that paper rose to par automatically. Yet, in 1878 the Bland Act was passed, requiring the purchase and coinage of not less than two million dollars worth of silver every month. The Treasury, too, had accumulated a considerable amount of gold, and it was decided that the mints were to be open to the free coinage of that metal. The currency, whether metallic or paper, was thus on the point of being enormously increased; and yet, with this increase plainly in view, the paper money rose to par. It is plain that this was due neither to supply nor to demand, but was simply the result of a change of opinion. Again, the gold premium in Buenos Ayres rose in the autumn of 1891 to about 360. It fell in the autumn of the following year to about 165, and it has risen this year to over 260. There was no such contraction of the currency between the autumn of 1891 and that of 1892 as would account for so extraordinary a rise in the value of the paper money as actually took place, nor has there been since such an increase in the currency as would explain the fall. The movements have been due simply to opinion. The peaceful election of President Saenz-Pena and his quiet installation in office filled everyone at home and abroad'with hopes that the Confederation was entering upon a period of recuperation and prosperity. The political troubles since have disappointed those hopes and inspired grave fears; and the paper money has fluctuated in value with the hopes and the fears. Once more, while the Sherman Bill was under discussion in the United States Congress, there sprang up a great speculation in silver, and the price rose from about 43⅝d. per ounce to 54⅝d. per ounce, a rise of about 25½ per cent. The Bill became law in July, 1890, and actually came into force in the middle of August. The highest price touched by silver was in the first week of September, when the Act was in operation not quite three weeks. Clearly the rise was purely speculative—was due, that is, not to supply nor to demand, but to the belief of the speculators that the value of the metal was about to be maintained at a very high figure. From the first week of September, 1890, the price fell with few interruptions, until, in June of this year, it touched 30d., from which it has since recovered to about 34d. The fall was as manifestly due to opinion as the previous rise had been. The world, generally, made up its mind that the United States experiment must fail, and, therefore, in spite of all the American Government, the great mining companies, and the speculators could do, the silver market broke in the way we have seen. Still another instance is furnished by the rapid fluctuations in the price of wheat during 1891. Between the end of March and the middle of May there was a rise of about 9s. per quarter. Then there was a sharp fall, then there was another rise—the highest point being reached in September. There was another fall and another rise, but after December, 1891, there was a steady downward movement. These fluctuations are explained by varying news as to the prospects of the crops all over Europe, and especially in Russia, during the summer ; in the autumn and in the winter, by the reports concerning the quantities of grain available for export. Here, again, we see values regulated neither by supply nor by demand, but simply by opinion, every change of the latter being quickly reflected in the markets. Lastly, every careful observer of the Stock Exchange is aware that a skilful operator may, and frequently does, buy a great amount of stock without very much influencing prices. Usually he employs more than one broker, and the brokers so cleverly buy and sell that the market is confused, and is not quite sure what is going on. An unusually great demand may in this way be satisfied without such a change in prices as would seem inevitable. And, similarly, a very large amount may be sold without such a fall as beforehand would certainly be predicted. On the other hand, if it is noised abroad that a great capitalist whose judgment is respected is buying largely there is sure to be a rapid rise.

THE THEORY OF VALUE.*—II.

THE standard works on Political Economy say that everything which is capable of gratifying a human desire, and which is exchangeable, has value. Has it? To enjoy the good things of this world without working for them is a strong human desire; and power to compel men and women to work for one without paying them gratifies that desire, while obviously men and women are exchangeable. Therefore, until a recent date in the history of the world, men and women were almost everywhere made slaves, and were bought and sold just as cattle are. But now, outside of the possessions of Spain, slavery is prohibited in all civilised countries. Thus we see that men and women had value in the economic sense of the word almost everywhere formerly, and that they have not value now in most parts of the civilised world. Again, in primitive communities land has not value in the full economic sense. It belongs to the clan, or tribe, or village community, and the members of the community have the usufruct of it. Often it is redistributed amongst them at stated intervals; but it is not marketable, it is not private property, it cannot be alienated either by the community or by the clan chief. The land, in fact, is occupied but it is not owned, properly speaking; it is rather like the air and the sunshine. In still more primitive states it is not even fully occupied. The nomad tribes

---

* From *The Statist* of November 18, 1893.

move according to the seasons from district to district. Mostly the nomad tribes return to the same places at the respective seasons, but they have not our conception of ownership. Their claim more nearly resembles that of the dog to his special corner. To take a third instance, the right of presentation to an ecclesiastical benefice is property in England; it has value and is bought and sold. In many other countries to attempt to deal in it would be considered sacriligeous. Up to a few years ago, furthermore, commissions in the army were, in the same way, property, and were openly dealt in; they have now ceased to have value. Lastly, all kinds of public offices, including Judgeships, formerly were matters of bargain. But traffic in such offices now is considered corruption, and is rigorously put down in civilised countries. Surely that is not an accurate definition which is true in Cuba and not in England, which was true formerly, but is so no longer, which is true in the present, and was not true in the past. A definition to be worth anything must be true always and everywhere; a single exception proves it to be incomplete. The generally accepted definition of value, then, is clearly incomplete. It leaves out something essential, something that differentiates value from mere utility, as well as from utility combined with exchangeability. I venture to think that economists generally have misapprehended the real nature of value because they have not paid due attention to its subjectivity.

Every competent economist, of course, would admit, if challenged, that value is subjective, that it is not a something inherent in external objects, but is attributed to them by us. But none of them give importance to the fact, and most would expressly deny that importance is due to it. They would say that what we call the qualities of external objects are all subjective, and so, no doubt, they are, but not in the same sense. For example, the hardness of a piece of iron is always the same, provided heat is not applied; it does not vary with our moods. But the value of a thing does vary with our moods. A caprice of fashion lifts a thing into sudden estimation and drops it just as quickly—a fabric, a garment, a colour, a piece of old china. The crinoline and the chignon of a few years ago are not of the same value now as they were then, simply because fashion has changed. It has been well said that the history of civilisation is the history of inventions, and the history of inventions from one point of view is the history of the introducing and discarding of values. There are countless multitudes of things that were once of high value and that now have little or none. The armour, so indispensable to the mail-clad knight, and the manufacture of which gave employment to so large a proportion of people in the middle ages, is now only an object of curiosity. The windmills so frequent in English scenery represent a dying industry. The domestic spinning-wheel, though still to be found in the cottages of remote districts and in backward countries, has gone completely out of use everywhere else. These are but a few illustrations; there are countless others, some of which we are acquainted with only in museums, some are known solely through the writings of ancient authors, the memory of others is so completely lost that when a specimen is dug up by the explorer of some ruin its use can only be guessed at. It may be said that invention creates or destroys demand, and that it is demand which determines value. Even if that could be admitted—and it cannot—it would be beside the point. If we were to stop at saying that a man is killed by poison, or a knife, or a bullet, we should never get at the murderer; nay, there would be no such crime as murder on the Statute Book, and consequently there would be no such criminal as a murderer to be punished.

Invention, however, is only one of the causes that bring about a great change in values. Closer communication between distant countries—itself a consequence of invention—has likewise a potent influence. Tea and coffee were unknown in Europe a few centuries ago; now they have become necessaries of life, and have driven beer from the breakfast table. The change of habit has added immensely to the value of land and the remuneration of labour in the Far East and in South America, while it has kept down the value of land in Europe by lessening the consumption of European grain. A change in religion has likewise a considerable influence on values. The Reformation, for example, destroyed the trade in priests' vestments in this country; the High Church revival has brought it to life again. In an exclusively Hindoo community the flesh of the cow has no value for human food; nor has the flesh of the pig in an exclusively Mohammedan; and it is obvious that if ever vegetarianism becomes universal, meat will cease to have value, and not only will the butcher's trade be destroyed, but so also will all the industries dependent upon the fattening of cattle. In the same way, if ever total abstinence becomes universal, the liquor trade will be destroyed, and so will the cultivation of the vine and of grain for both brewing and distilling. The development of the moral sense likewise has a powerful influence upon values. I have referred above to slavery. Last century the Government of this country was thought to have gained a great public advantage when it acquired a share in the Spanish slave trade; in the present century our Government induced other States to join with it in keeping a squadron on the West Coast of Africa to put down the slave trade. In a single century the change of opinion has converted what used to be regarded not only as a gainful but as a legitimate traffic into a crime against humanity. In short, every great change in the habits of a people must have a most important bearing upon all values.

If the foregoing argument be admitted, it follows that in the same country the same thing may have very great value at one time, and little or no value at another. In like manner at the same time the same thing may have great value in one country and little or none in another. An adult European would hardly stoop to pick up the glass beads which are so highly prized in Central Africa that the natives will exchange ivory for them. On the other hand, the ivory is valued so little in Africa that it is freely exchanged for the glass beads; and yet it is in such estimation in Europe that men brave all the dangers of African travel to obtain it. Similarly the cowries which pass as money in India have no

value here. Even amongst civilised countries the same phenomenon may be observed. For example, a bank note which is
everywhere taken at home at its nominal value has no currency
abroad. Value, then, varies with our moods, varies likewise from
country to country, varies, moreover, in the same country from
time to time ; and yet the thing to which it is attributed has always, and everywhere, precisely the same physical qualities. It is
equally exchangeable and equally capable of gratifying a human
desire. Clearly, therefore, everything which is exchangeable and
which can gratify a human desire has not value, for things that
have had value have lost it, and things that had no value have acquired it. Equally, clearly, value is a function of opinion, and of
opinion, too, which varies from time to time, and from place to
place. I venture to submit, then, that value is given by the local
and temporary opinion which selects from the mass of exchangeable things capable of gratifying human desires those best suited
at the time and in the stage of civilisation to do so.

THE THEORY OF VALUE.*—III.

FROM the foregoing it appears that it is not utility which gives
value, nor utility combined with exchangeability, but it is the local
and temporary opinion which selects from the mass of exchangeable
things capable of gratifying human desires those best suited at the
time and in the stage of civilisation to do so. Selection by opinion
is the first essential of value. The second essential is inaccurately
and deceptively described by English economists as exchangeability. I submit that exchangeability does not confer value ;
and that neither does scarcity. What is really essential is
appropriability—if I may coin a word—or the capability of being
appropriated. It is quite true that what is exchangeable is

---

* From *The Statist* of January 13, 1894.

usually also capable of being appropriated. But it is not the mere fact that it is exchangeable which confers value ; it is the fact that the thing has been appropriated and that nobody can, therefore, get it without giving something for it—at least peacefully and according to law. So again, it is not scarcity which confers value, though the Austrian School, which is in absurd estimation in this country just now, has revived that exploded fallacy.

The most useful of all things—or at all events the most indispensable of all things—are, first, the air we breathe, without which we could not live for a single moment ; secondly, the rays of the sun, which give to our earth the warmth and light, without which life—at all events as we know it on this planet—would be impossible ; and thirdly, the rain which gives the moisture, without which there could be no vegetation, and consequently there could be no food to maintain life. But of these three indispensably useful things two do *not* exist in superfluity. Rain, for example, is often so scarce that vast tracts of the earth suffer from drought, and in the more backward countries the drought leads to disastrous famines. Again, sunshine is certainly not in superfluity either in the Arctic or Antarctic circles, in both of which there is no sunshine for half the year. On the other hand, limitation of supply, or scarcity, or whatever other word be preferred, does not necessarily confer value. Up to the other day our turnpike roads had value. Nobody could use them, at least with a vehicle of any kind, without paying for their use. But the turnpike roads have now been freed, and they no longer have value. Yet the turnpike roads are just as limited in supply now as they were before. It is not limitation of supply, or scarcity, then, which confers value, and it is not superfluity which prevents value from arising. What really does give value is appropriation. Happily no man, or class of men, or community, has ever yet been able to appropriate the sunshine or the rain, and to say to all the rest of mankind that they shall enjoy neither unless they pay for the privilege. Therefore the rain and the sunshine are free, not because they are unlimited, as has been just shown, but because nobody has ever yet been able to appropriate them. If the rain-making inventions, of which we have heard so much of late, ever succeed, doubtless the Governments of each country in the world will take to themselves the right of working the inventions, and the Governments then will be able to charge for rain, which thereby will acquire value. But rain will not, if the idea is realised, be more limited in supply or scarcer than it is now; on the contrary, it will be much more plentiful, for the Governments will be able to bring on rain when they please.

In the standard English works on Political Economy, the power of appropriation, which so largely contributes to confer value, is called exchangeability. It is by no means a good word, but it is so well established in the science now that it is not worth while to try to displace it. Those, however, who really wish thoroughly to understand the meaning of the word " value," must clearly bear in mind that it is the power of appropriating to one's own use and of excluding others from the use of a thing, which mainly contributes to give value. Of course the thing must have utility. If a person cared nothing for the thing or what it could do for him, he would not take the trouble to appropriate it. In a sense, then, utility is the basis of value. We should ourselves prefer to say that utility " conditions " value, and that even with utility value cannot exist unless a man can appropriate the thing to his own use. Having appropriated it, he excludes other people from its use, and if other people wish to share in the use, or to acquire the use of it, he can insist upon a price or a return, or an equivalent. Value, therefore, from one point of view, is rather the opposite to utility than a consequence of it. So far as those who are excluded are concerned, at all events, value is a charge. To those who have appropriated a thing, on the other hand, value is a benefit. But, of course, as already said, the thing must have utility, must contribute, that is, in some way or other to the gratification of a human desire, or nobody would care either to appropriate it or to purchase or hire it from others who had already appropriated it. Utility, therefore, is a condition of value, but it no more enters into the definition of value than existence. A thing cannot have value unless it exists, and precisely similarly it cannot have value unless it has utility. But mere utility, no matter how combined with other things, does not confer value. What first and mainly contributes to confer it is the opinion local and temporary that a thing is best qualified to gratify a human desire. And what in the second place contributes to confer it is not scarcity, but capability of appropriation.

THE THEORY OF VALUE.*—IV.

VALUE being given by the local and contemporary opinion as to the things best suited to gratify human desires, relative values are determined by the local and temporary estimates of the probabilities of obtaining those things in sufficient quantities; and the estimates are in each case themselves determined mainly by the strength of the desires and by the efficiency of labour. At first sight it may seem contrary to all experience to say that the strength of the desires is one of the main factors in determining the values of things. If it were, it may be objected, food and drink would be the dearest and diamonds among the cheapest of commodities. But a fuller consideration of the phenomena, not in the more advanced countries only but in all communities throughout the world's history, will, I venture to think, change the first impression. In the most primitive communities we know of, men pass their time in gorging, sleeping, hunting, fishing, and fighting. The efficiency of labour is at the lowest, production is likewise at the smallest, and consequently there are exceedingly few values. Hardly anything but those which gratify the strongest desires is

---

* From *The Statist* of December 9, 1893.

produced. Even in those communities there is not a complete equality of condition. There is, therefore, some little superfluity. The strong are able to gratify desires not literally urgent at the expense of the weak ; but speaking broadly it is true to say that few things have value which do not gratify the strongest desires. Every advance in knowledge increases the efficiency of labour, and therefore each step forward in civilisation leads to a larger production, and so brings into existence new values. As soon as anyone obtains more than he needs for gratifying his strongest desires, he is able to gratify weaker desires, and in so doing he introduces new values. Everyone knows how invention cheapens commodities. This is generally explained by saying that the cost of production is reduced by the invention. I venture to submit that that is a misleading way of putting the facts. It turns attention from what is essential to what is accidental. What is essential is that human desires are the same always and every-where, in the savage and in the civilised man, that most of them cannot be gratified in the savage state owing to the inefficiency of labour, that everything which makes labour more efficient enables new desires to be gratified, and that it is by making labour more efficient that invention cheapens commodities—not by merely reducing the cost of production. It is quite conceivable that a reduction in the cost of production might be so great as to cause the thing produced to lose value altogether. At the present time, for example, diamonds are prized by the very wealthy because of their rarity and cost, because, in short, very few can wear them. If they could be produced so cheaply that the seamstress earning only a few pence a day could buy them, they would lose all attraction for the wealthy, and probably before long would cease altogether to be worn. But an invention that cheapens a thing, the desire for which is constant, acts altogether differently. Every invention, for example, that has cheapened calico has brought it into more extensive use. The reducing of the cost of production, then, is a mere accident ; what is essential is the bring-ing within the reach of the many of a thing desired by them all.

The strong desires affect values in two ways. Directly they tend to make the things desired dear. In the savage state, as already said, they prevent nearly all other things from having value. But indirectly they tend to cheapen things, for they compel the vast majority of mankind to employ themselves in producing those things; and as the efficiency of labour increases, the things become more and more abundant, and so it becomes possible for larger and larger proportions to gratify weaker desires. In every stage of civilisation, however, the strong desires limit values. Even in our own country the very poor are in a position to gratify only the strong desires, and for them, consequently, there are few values. The class immediately above them in wealth is in a position to gratify a few more desires, and so on up to the very wealthy whose surplus is exceedingly large, and for whom, therefore, there are innumerable values. Economists generally have too much confined their attention to the great commercial communities, and even in observing these they have not remarked sufficiently upon the different economic conditions of the different classes ; otherwise they could not have so greatly mistaken the nature of value, and especially they could not have so utterly overlooked the important part played by the strength of the desires. As civilisation advances and the efficiency of labour increases it becomes more and more easy to gratify the stronger desires. Then for a proportion of the population it becomes possible to gratify weaker desires, and ultimately the proportion so circumstanced increases, while at the same time the most favourably-placed portion of the population is constantly being enabled to gratify new desires. But this is only because the strong desires compel the vast majority to devote themselves to producing what will gratify those desires. Here in England the majority of the population is employed in rendering services and in manufactures. In a few other rich countries a very large proportion is similarly employed; but over far the larger part of the world the vast majority is employed in producing the necessaries of life. The strong desires, then, operate in two ways. Directly they tend to make the things that gratify them dear ; indirectly they tend to make them cheap by compelling so many to produce them. Owing to this latter circumstance, and to the increasing efficiency of labour, the things that gratify the strong desires tend to become cheaper and cheaper as time goes on ; that is to say, experience enables people generally to estimate that they will be able to gratify the strong desires at such and such an outlay of their labour, and then the weaker desires come into play. Usually in the more advanced countries, the things that gratify the weaker desires are the dearest ; firstly, because only a small proportion of mankind is engaged in producing them. The very strong desires, as already said, compel the vast majority to engage in producing the things that satisfy those desires. The small minority left can be tempted into producing other things only by the offer of higher and higher remuneration. Here again it will be seen that the strength of the desires continues to operate. Secondly, it is usually found that in regard to the things which gratify the weaker desires, we have little or no means of increasing the efficiency of labour. A new invention may enable us to produce a hundred times as much of a given manufacture at a smaller cost ; but no invention has ever taught us to turn a dullard into a genius.

The estimate of the probability of obtaining a thing, then, is mainly determined by the strength of the desire and the efficiency of labour. It is a mere estimate, constantly in a state of flux, and is modified by every circumstance that seems likely to influence the efficiency of labour whether permanently or temporarily—an invention, or reported invention, a flood, a drought, a war, or a convulsion of nature. Where the efficiency of labour is low, or where it may be neutralised by natural causes, the strong desires every now and then assert themselves and almost annihilate the values of the things which gratify the weaker desires. For example, in a great famine almost everything but food loses value. In the more advanced countries of the modern world we draw our supplies from all parts of the earth, and in normal times, therefore, do not know what the fear of famine is. But even in the most civilised countries in abnormal times the values of most things may practically be destroyed. In the latter days of the siege of Paris, for example, what would not a man have given for a good dinner?

THE DEPRECIATION OF SILVER.*—V.

A MOST signal instance of the influence exercised by opinion upon value, because of the widespread and unforeseen consequences, both political and economic, likely to ensue, is presented by the depreciation of silver during the past twenty years. In the course of the last century, gold gradually came into very general use in this country—so much so, that when the resumption of specie payments was carried out after the great war against Napoleon, it was decided to adopt the single gold standard and to use silver, not as a standard of value in future, but merely as subsidiary coin. Gradually the preference thus shown for gold in the United Kingdom began to exhibit itself upon the Continent likewise. When the great gold discoveries in California and Australia took place, the preference was quickened. Silver was exported largely from the Continent to the Far East, and gold was imported to take its place. The richer nations began to see that they could establish a gold standard without very serious inconvenience to themselves. The reader will recollect that, at the International Monetary Conference held in Paris, in 1867, a very strong inclination was shown to adopt the single gold standard; and no doubt can be entertained by anyone who studies the matter carefully and without prejudice, that, if the Franco-German war had not

---

* From *The Statist* of December 9, 1893.

taken place, France would very soon have given up bimetallism and adopted the single gold standard. The war, however, did break out, with the result that Germany imposed an indemnity of 200 millions sterling upon France. That indemnity enabled Germany to establish a uniform monetary system throughout the whole Empire, instead of the separate systems that the several States had had previously; and in doing this the German authorities demonetised silver and adopted the gold standard. In 1873 the United States also demonetised silver and adopted gold. In 1878, it is true, it partly retraced the step by the Bland Act, and it made a further backward step in 1890 by the Sherman Act. But this year, as everybody knows, not only the coinage of silver but the purchase of the metal has been completely stopped. As a consequence of the sales of the German silver, the nations of the Latin Union very soon closed their mints against the metal. Last year Austria-Hungary adopted the single gold standard, rejecting silver, and this year India has closed her mints against silver. Practically, therefore, silver has ceased to be a standard of value throughout the civilised world; and this has simply and solely been in consequence of a slow change in the opinion of the world.

The Bimetallists deny that the world has changed its opinion in the manner here stated. They allege that the whole change has been made by the Governments, and not by the people. If the Bimetallists mean by this that the great body of the people have not troubled their heads about the matter, I am not inclined to dispute the statement. But if they mean that the Governments acted on their own initiative, and not because of the impulse given by the commercial classes, then I entirely disagree with them. Governments, however meddlesome in other matters, are always very careful how they interfere in business arrangements, and especially in monetary arrangements, and they do interfere finally only because they are compelled to do so by the pressure of the classes specially interested. The change has taken place, then, because there has been a complete revolution in the opinion of the banking, the financial, and the leading commercial classes throughout the civilised world. And the remarkable thing is that the advantage possessed by gold over silver does not seem great enough to bring about so extraordinary a revolution. There is no doubt, of course, that for international payments gold is a much more convenient metal than silver. Even twenty years ago the Latin Union rated silver to gold in the proportion of 15½ to 1; and it needs no argument to show that it is cheaper, easier, and more convenient to send abroad a ton of gold than fifteen-and-a-half tons of silver. But, after all, the people who have to make large gold international payments are not very numerous, and they have not influence enough to compel Governments to change their monetary systems. If, however, we look at the home business, it is difficult to see why men have so generally preferred gold to silver. No doubt it saves time and labour, both for bankers and for large employers of labour, to handle gold rather than silver. But if the large employers of labour save expense in paying in sovereigns and half-sovereigns as far as they can, and if the bankers likewise save expense, it is clearly far from convenient for the working classes. The very first thing an ordinary working man has to do when he is paid a sovereign is to change it into silver, for it is rarely that a single payment of his amounts to a pound sterling. If, then, gold is more convenient for the large employers and bankers, it is less convenient for the working classes. And besides, all the advantages of a gold circulation would be equally secured if one-pound notes were used instead of sovereigns. So again it may be said that it is much handier to carry about a few sovereigns than to load one's self with an equivalent value in silver. But, in the first place, it is not necessary to load one's self with silver, for notes would do just as well; and in the second place, very few men carry either gold or silver to any considerable amount about with them; what payments they have to make they usually make by means of cheques. The preference, then, is purely sentimental. It is just like the preference for sovereigns over one-pound notes in England, and it is not very easily explained by any real advantage that gold has over silver. But the fact remains, nevertheless, that one civilised country after another has demonetised silver and adopted the gold standard ; and there is no answer to that.

This is the real and conclusive argument against the Bimetallists. For over a century the world has been moving country by country steadily in one direction, and it is useless to talk of turning it back. No Government can act in opposition to the public opinion of its own subjects. It may be said that the public opinion is that of the wealthy, and consequently of a comparatively small class, that the multitude may be educated and may compel the wealthy to submit. That I hold to be utterly impossible, for no Government, however autocratic it may be, can prevent people from contracting to be paid in any manner they may choose to arrange. Opinion has settled the matter, and opinion is sovereign not only in matters of policy but in matters of economy also. If the Bimetallists choose to say that opinion is wrong, that the world is ill-advised, that the whole movement has been a mistake, they, of course, have a perfect right to their view of the matter. But they do not alter the facts, and they cannot alter the facts by any amount of preaching.

THE BIMETALLIC CONFERENCE.*

THE Bimetallic Conference was undoubtedly an imposing demonstration, and shows clearly that the movement is gathering strength. But for all that, it is doomed to failure. Mr. Balfour and Mr. Courtney attempted to show that bimetallism is practicable. I maintain, on the contrary, that it is impracticable, and that

* From *The Statist* of May 5, 1894

none of their arguments touch the heart of the matter. It is impossible by any arrangement whatsoever to maintain indefinitely a fixed ratio of value between any two commodities, let them be metals, or let them be what you please. And it is noteworthy that no speaker ventured to offer any detailed scheme showing how the idea could be carried out. No one ventured either to take a vote on the ratio to be fixed. The whole history of the modern world is full of examples of Governments endeavouring, and endeavouring in vain, to establish such a ratio. But our bimetallic friends point to the success of French bimetallism as indisputably confirming their argument. I deny in the clearest and most emphatic manner that it does anything of the kind. French bimetallism and the Latin Union in their full integrity existed for barely seventy years. At the end of that time bimetallism broke down hopelessly. During the seventy years I admit freely that the variations in the value of gold and silver were not very great; but that was due, not to the existence of the Latin Union, but to entirely different causes. The first of these great causes was that the supply of silver did not very greatly increase or decrease, and that the consumption of the metal also did not very greatly increase or decrease. There was a fairly constant relation between the supply and the demand. The second of the great causes was that there was a remarkable and almost phenomenal change in the supply of gold. About 1848 the great gold discoveries in California and Australia led to an extraordinary increase in the supply of gold. About twelve years later, the civil war in the United States caused a suspension of specie payments, and put an end practically to the consumption of gold as money in that country. Then the Franco-German war led to the suspension of gold coinage for some years in France. Thus there was an enormous increase in the supply of gold, and a very considerable decrease in the consumption of that metal. The real reason, then, why the conventional ratio between gold and silver was maintained from the beginning of the century until the demonetisation of silver by Germany was, not that France maintained a bimetallic system, but that there was no great increase or decrease either in the supply or the consumption of silver, and that there was an extraordinary increase in the supply of gold, and a very considerable decrease in the consumption of that metal. The facts being so—and I challenge any fair-minded man to dispute them—I submit that the Latin Union proves nothing as to the practicability of bimetallism.

My next point is that the nations cannot be induced to enter into a Bimetallic Convention. In spite of the conversion of a few weather-cocks like Mr. Courtney, who are stable in nothing, the real scientific opinion of this country is against bimetallism, and will continue to be opposed to it. Further, it seems incredible that Germany, Austria-Hungary, and Russia will throw away the fruits of all their sacrifices and all their efforts to get together a great gold reserve. Ever since the middle of the century the great nations of the Continent have been feverishly preparing themselves for war, and one of the methods of preparation is to accumulate a vast reserve of gold. Are we as reasonable men to be asked to believe that the great military powers will throw away the advantage which the possession of so much gold gives them, and thus waste the exertions and the sacrifices of so many years? Suppose that the Governments could be induced, that they were blind enough to give up a real advantage in pursuit of a shadow, how could a Bimetallic League be maintained? No country could bind itself for ever-and-a-day to maintain bimetallism. Even the Latin Union concluded its Convention only for a definite number of years. But if the nations were to form a Bimetallic Union for ten, or even for twenty years, what assurance would there be that some of them would not break away at the end of that time, as so many of them have broken away from commercial conventions? And if the Union were to be formed, and were to be compulsory only for a few years, is it reasonable to suppose that bankers would accept depreciated silver instead of the appreciated gold to which they are entitled? In *The Statist* I have over and over again shown, when predicting last year's great

currency crisis in the United States, that the banks throughout the Union had steadily refused to hold either silver or silver
certificates in their reserves to any extent worth speaking of, and
that it was the refusal of the banks to treat silver as the equivalent
of gold, and hold it in their reserves, which condemned the
American experiment to inevitable and certain failure. The event
has proved me right; and I predict now with equal confidence
that if every great nation in the world were to enter into a
Bimetallic Convention for a definite number of years, bankers in
the great commercial centres, and especially the bankers of
London, would refuse to accept silver and hold it in their reserves,
and the Convention would break down as certainly as the American experiment failed.

There is one other point never to be lost sight of in the discussion of this matter. It is a point that was dwelt upon at great
length and with much fulness by Mr. Gladstone in the House of
Commons, and it is one that bears to be repeated again and again.
It is the real interest of this country in the question. England
is admittedly and demonstrably the greatest creditor country in
the world. She has lent immense sums to most other countries ;
she has invested immense sums in North and South America, in
India, in our Colonies, and elsewhere in industrial enterprise of
every kind ; and she has contracted for the most part when lend-
ing that she is to be paid both principal and interest in gold. Is
it conceivable that the investors of this country would without
compensation and for no advantage to themselves waive the condition they have imposed as to repayment in gold, and consent to
receive depreciated silver instead ? Does any sensible man of
business believe that Englishmen will do this ? Or would any
conscientious man, if called upon professionally to advise, recommend a client to do so ? I unhesitatingly answer both questions
in the negative. Bimetallists airily reply that there would be no
loss, because silver and gold would be receivable everywhere at
the fixed ratio. To which I rejoin that that is a pure assumption ; that in my opinion silver and gold would not be everywhere

receivable at a fixed ratio ; that bankers, especially in the City of London, would refuse to receive silver, would contract to be paid in gold, would decline to hold silver in their reserves, and would insist upon the condition which was so frequently introduced in America while the Bland and the Sherman Acts were on the Statute Book—for repayment in gold, and in gold alone. Besides, there would always remain the fact that the Convention would be terminable, that any party to it might withdraw at a certain time by giving a certain notice. Therefore, there would be doubts whether the Convention would last, and every careful man, bearing this in mind, would look askant upon silver, and would attach a greater value than ever before to his gold contracts. Suppose that English investors were so unwise as to waive their gold contracts and to accept payment in silver ; and suppose, as a consequence, that the United States and France were to succeed in pouring into this country a large proportion of the unusable silver now held by them, is it not at least possible that both the United States and France would chuckle at our folly, and would give notice to terminate the Convention when the time came ? But if English investors refused to waive their gold contracts, and if English bankers declined to hold silver in their reserves, what would become of a Bimetallic Convention, supposing the Governments were foolish enough to enter into one ? Clearly, bimetallists have not thought out the subject. I would respectfully recommend them, before airing their views again, to more deeply study the obstacles to be overcome.

the ratio fixed between silver and gold. But that was due to an extraordinary combination of circumstances. Firstly, the production and the consumption of silver neither increased nor decreased very much, while there was an almost unprecedented increase in the production of gold, and there was an extraordinary decrease in its consumption owing to the suspension of specie payments first in the United States and then in France. Thus to maintain the French ratio, not only was it necessary that all the countries then using silver should remain silver-using ; it was also necessary that an almost unprecedented production of gold should take place. Will any reasonable man say that we have a right to expect in the future for seventy years, or even for ten years, such an unlooked-for combination of favourable influences ?

It is one of the surprising things connected with the bimetallic agitation that so many intelligent men fail to see the real bearing of this appeal to the Latin Union experience. To my mind, it appeals as clear as the sun at noonday that the appeal establishes in the most conclusive manner the impracticability of bimetallism. With all the favourable influences during the first three-quarters of the century, the very instant there was a disturbance of the old arrangements, the whole house of cards toppled together. Germany having imposed upon France what was then thought a crushing war indemnity, proceeded immediately to demonetise silver and to adopt the single gold standard. As soon as she began to carry out her policy, the ratio established between gold and silver came to an end, and the divergence has been widening every year since. Very quickly the Latin Union suspended the coinage of standard silver pieces, which has never been resumed since ; and one after another a number of States have followed the example of Germany, so that practically at the present moment silver has ceased to be a standard of value throughout the civilised world. Is not this as strong a proof as could be produced that only by an accident could bimetallism be made to work even for a few years, and that it is in the highest degree improbable that any combination of accidents would make it work for many years ? But the

Bimetallists will not see the impracticability of their proposals, nor will they recognise that Governments cannot be persuaded to fall in with their plans. It has been pointed out above how very unlikely it is that Germany, Austria-Hungary, and Russia—not to mention our own country—will adopt bimetallism. But the Bimetallists insist that if our own country will lead the way all the other countries will immediately follow. This is, no doubt, paying us a very high compliment, but we are not prepared to accept the flattery. Again, the Bimetallists will not see that even if the Governments could be persuaded there would still be the bankers to deal with, that bimetallism could be adopted only for a definite number of years ; that therefore, the danger would remain that it might come to an end by-and-by, and as long as a fear of that kind existed people would cling more fondly than ever to their gold contracts. From every point of view, then, bimetallism is impossible. It could not be made to work if adopted. It will not be adopted because the Governments cannot be persuaded to fall in with it, and if the Governments could be persuaded it would be rejected by the bankers, and still more decisively would it be rejected by investors.

The truth is that the Bimetallists are trying to turn back the hands of the clock of time. Once silver was the standard all over Europe. Gradually gold came more and more into use. During the present century gold has been adopted as the sole standard—first in England and afterwards by most other countries. The result is, as already said, that although silver is a nominal standard in the Latin Union and in the United States, it can no longer be freely coined, and, therefore, has ceased to be a real standard throughout the civilised world. All this may be regrettable but life is too full of living problems to allow us time for vain regrets. We have to accept accomplished facts, and it is an accomplished fact that opinion has pronounced decisively against silver and for gold. And it must never be forgotten that in matters economic, opinion is as supreme as in matters political. Bimetallists will reply, no doubt, that they have set themselves to convert opinion. By all means let them endeavour to do so, but it is not rash to predict that they will fail. A *cult* once dead can never be brought to life again. Bimetallism was a convenient and a natural stage in economic development. Practically, as already said, silver was at one time the money of Europe. Then, when gold came into use, it was almost inevitable that bimetallism should be adopted. But it was adopted only for the purpose of facilitating the passage from the use of silver to the use of gold. The passage has now been effected, and it is certain that the nations will not go back to an obsolete system belonging to a period of transition that has come to an end.

THE LATIN UNION ARGUMENT.*—I.

As the part really played by French bimetallism—which is often conveniently, though erroneously called the Latin Union—seems to me to be widely misunderstood by Monometallists as well as by Bimetallists, I think it well to set the facts in the light in which they appear to my mind. French bimetallism was established in 1803, and in its full integrity it lasted until 1873, when it utterly broke down, and free coinage had to be suspended. During the whole of that time it is admitted by all who have studied the subject that the ratio between gold and silver was not strictly maintained, that there was always a premium on one or other of the metals—generally about one per cent., sometimes two per cent., and occasionally even three per cent. Intelligent Bimetallists reply that they do not pretend that a ratio can be strictly maintained. All they assert is that the variations can be kept within

---

* From *The Statist* of May 19, 1894.

very narrow limits, and practically, therefore, stable values assured. I venture to rejoin that they have not read the facts aright. French bimetallism did not prevent wide variations. What it really did was to facilitate the transition from one monetary system to another ; in other words, to enable Europe to pass with less friction than perhaps would otherwise have been possible from the silver to the gold standard. Bimetallism in its very nature, as I hope to show, is a temporary makeshift which is useful in a transition period, but which must inevitably break down when a certain evolution has been accomplished.

During the great revolutionary war specie payments were suspended in this country. But in 1821 they were resumed, and the single gold standard was deliberately adopted, silver being made legal tender only up to 40s., for the rest being used only for subsidiary or token coin. The English demand for gold to carry out this plan was very great, and it practically resulted in draining France of that metal. While this was going on the revolt of the Spanish Colonies in America and the anarchy that ensued interfered with mining and seriously lessened the output of gold. When the change of standard in this country was completed the United States began a somewhat similar change. The United States was bimetallic in theory ; but originally it fixed the ratio between silver and gold at 15 to 1. By that ratio silver was overvalued, and the consequence was that gold flowed away from the United States, and silver remained the only money. In 1834 the United States changed the ratio from 15 to 1 to 16 to 1, when apparently silver was undervalued. Silver consequently flowed away from the United States and gold poured in. Thus up to 1848 the result of French bimetallism was to enable the United Kingdom, in the first place, to carry out the resumption of specie payments upon a gold standard ; secondly, to enable the United States to substitute a gold for a silver currency ; and thirdly, to leave the whole Continent with silver only. I do not apprehend that these facts will be disputed ; but it may be worth while to quote in support of them the following extract from the evidence given by N. M. Rothschild, Esq., before the Committee on the Bank of England Charter, July 24th, 1832 " Question : Does the demand for gold from France produce a scarcity of money in France ? Answer : No. Question : Why is that ? Answer : Because gold is generally in private hands ; it is merchandise there. Question : If there was a demand for silver from France, would not that produce a scarcity of money in France ? Answer : Certainly, because it is the coinage of the country." It will be seen from this brief historical survey that up to 1848 bimetallism was not operative in France, gold was not in circulation. It was mere merchandise, held only by private persons. What really determined the value of the metals was the reduction of the gold supply by the revolutions in Spanish America, the demand for gold for the United Kingdom and the United States, and the use of silver by all the rest of the world.

At the end of 1847, the Marquis of Dalhousie was sent out as Governor-General of India. During his reign he annexed the Punjaub, Nagpore, Sattera, Jhansi, Berar, Oudh, and part of Burmah, thus practically completing the Empire of India. Since then, with the brief interruption of the Mutiny, the British Peace has been successfully maintained. There has been an extraordinary development of the country, a great advance in prosperity, and a marvellous absorption of silver. On November 26, 1892, there was published in *The Statist* a Table showing the course of Indian trade during the thirty-three years ended with March, 1892. The statistics are most remarkable, and will be found valuable by all who care to understand the economic development of the world during the period to which they refer. According to that Table, India absorbed during the thirty-three years 230½ millions sterling of silver, being at the rate of just 7 millions sterling per annum. In 1843 the first Chinese war opened five ports to European trade, and the second and third wars extended European communication with China. In consequence there was a vast increase in the Chinese demand for silver. In 1856 Japan was opened up, still further increasing the demand for silver for the Far East.

The opening up of these two great empires, and the growth of our Australasian Colonies, gave vast commercial importance to Singapore, for which also there has been a great increase in the demand for silver. But while all this was going on there was no material increase in the production of silver. It is as certain as anything can be which has not happened that if the European consumption of silver had continued on the scale previous to 1848 there would have been an extraordinary rise in the value of silver; in short, silver would have gone up probably as much as gold has advanced during the past twenty years. But the European consumption of silver did not continue on the old scale. The construction of railways, which had been so active here at home in the Forties, was pushed rapidly forward on the Continent in the Fifties ; and that together with the steamship and the telegraph, the opening up of the Far East, the rapid settlement of the United States and other new countries, and the long peace, vastly stimulated the growth of wealth in France. As transactions became larger, gold was felt to be more serviceable as money than silver. Meanwhile, the discoveries of gold in California and Australia, following upon those in Russia, towards the end of the first half of the century, added during the next twenty years an enormous amount to the gold supply of the world ; and France, which up to 1848 had been a strictly silver-using country, though theoretically bimetallic, rapidly became a strictly gold-using country, though still theoretically bimetallic. In his famous report upon the payment of the war idemnity by France to Germany, M. Léon Say states that the coinage of gold in France from 1848 to 1871 inclusive exceeded 6,640 millions of francs, or somewhat more than 265½ millions sterling. If during this period France had demonetised silver as Germany did a little later, no doubt silver would• have fallen rapidly in spite of the extraordinary demand for the Far East, because, as I have often insisted, opinion is supreme in matters economic, and opinion would have been alarmed if France had demonetised one of the metals. But France, although she did not demonetise silver, practically disposed of her stock to the Far East, and replaced it with gold. Gold being cheaper owing to the new discoveries, and silver being dearer owing to the demand for the Far East, it was a profitable transaction. It was carried out quietly and the world was not alarmed. But the result was that the transition of Europe from the silver to the gold standard was so nearly completed that in 1867 a Conference was held in Paris which decided in favour of the adoption of gold as the sole standard of value. In all human probability, if the war with Germany had not broken out, that decision would have been acted upon ; silver would have been demonetised, and would have fallen in value, as it has fallen in value since. But the Franco-German war prevented France from carrying out her intention. Still, the fact that France got rid of her silver, substituted gold for it, and so prepared to adopt the single gold standard, proves, I venture to submit, as clearly as anything can, that French bi-metallism was merely a temporary expedient calculated to facilitate a transition from one monetary system to another, but containing in itself the germs of inevitable dissolution.

THE LATIN UNION ARGUMENT.*—II.

I HAVE now shown, I hope to the satisfaction of all who are open to conviction, that during the seventy years of its nominal existence in its integrity, French bimetallism was never operative. Up to 1848 France was a silver-using country; afterwards it became gold-using. In the earlier period gold was not in circulation, and the mere fact that by law the Mint was open to the free coinage of gold had no more effect upon the value of gold than the mere fact that the London Tavern is open to customers has upon the

---

* From *The Statist* of May 26, 1890.

value of provisions. What really determined the value of both gold and silver was, in the first place, the fact that until the discoveries in California and Australia, only two countries—the United Kingdom and United States—can be said to have been gold-using; all the rest were silver-using. After the discoveries in California and Australia, the enormous demand of France for gold, and of the Far East for silver, kept up the values of both metals. But the existence of bimetallism unquestionably was serviceable in a period of transition. It probably prevented a good deal of friction. In the meantime, it educated the world in a preference for gold. In 1821, our readers will recollect, the United Kingdom adopted the single gold standard. In 1834 the United States changed the existing ratio of silver to gold for the very purpose of getting the latter metal, and in 1867 France declared for the gold standard. Thus it is in the highest degree probable that if the Franco-German war had not broken out France would have demonetised silver, and the fall which we have since experienced would have taken place. But the Franco-German war prevented France from carrying out her idea and gave an opportunity to Germany to forestall her. Germany was so completely victorious that besides annexing Alsace and Lorraine, she imposed a money fine under the name of a war indemnity of 200 millions sterling upon France.

Germany immediately proceeded to reform her currency. In the interest of her own unity that was a politic measure. But, of course, she might have made her money uniform while preserving the single silver standard. She decided however to demonetise silver and to adopt the single gold standard. In doing this she applied an unerring test to the practicability and the worth of bimetallism. If it had stood the strain, then there would have been much force in the arguments of its advocates. As it did not stand the strain, but broke down instantly, it is proved to have been utterly worthless. It was a mere form without any real significance. The result of Germany's policy is instructive in another way. It not only shows, as I have been insisting, that bimetallism was unable to stand a serious strain, but it proves, as also I have been contending, that the world had been educated in a preference for gold ; for the instant Germany began to sell silver on a considerable scale, the nations of the Latin Union suspended the free coinage of the metal, showing clearly that they looked upon silver as an inferior metal and would not have it. It is often said that France was smarting under the disasters of the war, and that she closed her mints against silver for the very purpose of thwarting Germany. I do not dispute the assertion, but I submit that it proves completely my position. If France believed that silver was a superior metal to gold, she would have very gladly taken it from Germany, and thought that she had had a small part of the revenge she thirsted for. Even if France had believed that silver was as good a metal as gold for monetary purposes, what object could she have had in closing her mints? She would not have injured herself on the assumption by taking the silver, and she would not have injured Germany by refusing to take it, for silver would have remained in general estimation. But France knew that silver was not equal to gold for monetary purposes, and by closing her mints she did unquestionably make the German reform more difficult than it otherwise might have been. She caused the price of silver to fall more heavily than it otherwise would have done, and so she compelled the German Government after a while to stop its sales of the metal.

There is no doubt, then, that the action of France did obstruct the policy of Germany. The demonetisation of silver by Germany would, in any case, no doubt have caused a considerable fall in the value of the metal. The closing of the mints throughout the Latin Union caused the fall to be much greater than it would have been from the mere demonetisation by Germany. And the action of both was intensified by the demonetisation of silver by the United States in 1873. It has often been said that the Bill adopting the single gold standard in the United States was smuggled through Congress. For the purpose of this article it is not

D

worth while to enquire whether it was or was not so. The only thing of any real moment is that the single gold standard was adopted in 1873, that in spite of all the activity of the Silver Party since, the decision then taken has never been altered. The Bland Act, it is quite true, instructed the American Government to buy silver, and the Sherman Act increased the purchases largely. But neither Act opened the mints of the United States to the free coinage of silver ; and neither, therefore, returned to bimetallism in the full and proper sense. The Bland Act and the Sherman Act have both been repealed. It is now twenty-one years since the adoption of the single gold standard, and the decision then taken is still the law of the United States. Therefore I am justified in saying that the American people had become confirmed in their preference for gold. Thus the precedent set by the United Kingdom in 1821 has been followed by Germany and the United States—not to mention Austria-Hungary and other countries— explicitly and fully ; and it has been followed not quite so fully or explicitly by the nations of the Latin Union. After a trial of nearly a century the United States decided that bimetallism was a mistake, and adopted the single gold standard. After a trial of nearly three-quarters of a century, France came to the same conclusion. Can there be a stronger or more conclusive answer to the Bimetallists ? While the Western nations were thus one after another giving up the use of silver, the production of the metal began to increase immensely, and has gone on increasing every year since in an extraordinary way. This, of course, adds greatly to the depreciation.

But it is a mistake to say that the demonetisation of silver by so many countries, and the adoption of the single gold standard, together with the great increase in the production of silver—even if we add the marked decrease in the production of gold between 1870 and 1887—are the only causes of the unprecedented depreciation of silver. It may well be doubted whether there is not enough of gold for all practical purposes, were it not that so much of the metal is diverted from the use of the arts, and from employment as money by the great military governments. Ever since the Franco-German war, Europe has been preparing with breathless haste for another great struggle. Since the recovery of France— since, more particularly, the French army has once more gained a belief in itself—the preparations have been pushed forward more eagerly than ever. And amongst these various preparations, the hoarding of war treasure in gold has taken a very prominent place. Russia, as we know, has nearly 100 millions sterling in gold. Yet Russia does not even pretend that she is preparing for the resumption of specie payments. The world insists upon believing that Austria-Hungary decided to resume specie payments only to cover the accumulation of a vast stock of gold for military purposes. France holds in the Bank of France nearly 71 millions sterling in gold. Germany holds nearly 35 millions sterling in gold in the Imperial Bank, and 6 millions sterling more at Spandau. To all intents and purposes the gold held by Russia, Austria-Hungary, and the Bank of France is completely withdrawn from monetary employment. As regards Russia and Austria-Hungary, it is clear that it cannot be got at without the consent of the Governments; and the fact that bimetallism so far still exists in France, that the Bank of France can pay at pleasure either in silver or in gold, enables her to keep her hand upon the gold she has accumulated. The Imperial Bank of Germany is less free : but it is supported by the Government in discouraging gold withdrawals. The locking-up of these vast sums, then, has contributed powerfully to the depreciation of silver. And it would have raised the value of gold even if bimetallism had not been given up by France.

THE LATIN UNION ARGUMENT.*—III.

THE Bimetallists advocate bimetallism in the hope and belief that it would raise prices. On that point there can be no doubt, for in fact their strongest argument for bimetallism is that the demonetisation of silver has caused the crisis through which the world is now passing, and that there cannot be really prosperous trade until silver is restored to its old position as money; that, in short, the demonetisation of silver has caused the low prices. It would be only natural to expect that they would take some pains to assure themselves that high prices would result from the general adoption of bimetallism. But they have not done so: and I propose to point out now that, in fact, if bimetallism were to be adopted by the world immediately, it would not necessarily bring about high prices. High prices might result from the action of other causes and might be contemporaneous with bimetallism. But bimetallism alone would not cause high prices, and could not do so. Fortunately I am able to show this from a chart published the other day by Mr. Sauerbeck giving the course of prices from 1820 to the present time. Mr. Sauerbeck's statistics are based on the system of index numbers, and they represent the average of the prices of a large number of the principal commodities. There is much dispute as to the correctness of the index number system, but that need not be entered into here; the only point that concerns us is that Mr. Sauerbeck's industry, impartiality, and general correctness are universally recognised. His figures will be readily accepted by all parties, and they show clearly that the trend of prices ever since 1820 has been downwards—with, of course, very considerable, and sometimes very violent, fluctuations.

Mr. Sauerbeck takes the average of 1867–77 as 100. When the

---

* From *The Statist* of June 2, 1894.

average is above that there is a rise of prices; anything below that shows a fall. It may be mentioned that in 1800, taking the level of 1762 as the basis, the average was 141; that it fell to 110 three years later; that, with fluctuations, it gradually rose to 157 in 1809; then dropped hardly without a break to 115 in 1813; the following year fell again, until in 1816 it was as low as 91. I mention all this only to complete the history of the century. During the war there were so many disturbing influences at work that it would be altogether unfair and misleading to affect to draw general conclusions from the statistics of that time. But it may interest the reader, for all that, to know how prices actually moved during the war. After 1816 there was a rapid recovery, and in 1820 the average was 120, as compared with 100 for the period 1867-77. From this point Mr. Sauerbeck's statistics begin. The fall was almost immediate and continuous until 1822, when the average was barely 101. Then there was a very steady rise until 1825, when the average was as high as 117. Then there was a sharp fall, until in the next year the average was no more than 98½, and the fall continued till 1832, when the average was only 89. Then a rapid rise occurred, and in 1836 the average reached 102, and in 1839-40 it went to 103. A fall then began and went on until in 1844, when the average was as low as 83. After that there was a slow rise until 1847, when the average was 95; a sharp fall ensued till 1849, when the average was 74; fluctuations followed for a couple of years; then there was a rapid bound upwards till 1853, when the average reached 102; a slight decline followed next year, then steadiness for a couple of years, and then a further rise to 105 in 1857, after which there was a rapid fall till in 1859 the average was 91; then a rise up to 1864, when the average was again 105; a fall until in 1870 the average was 96; a rise till 1873, when the average was 111; and then after that an unbroken fall, until in 1879 the average was 83; a rise till 1881 when the average was 88; then a fall, until in 1887 the average was only 68; a rise in 1889 to an average of 72; and a fall last year to 68.

Thus starting in 1820 with the average of prices at 120, we find that there was a fall to 101 in 1822, and a recovery in 1825 to 117. But after 1825 there is an almost continuous decline till 1832, when the average was 89. In other words, in the period 1820–32 we have the highest average of prices, 120, and the lowest, 89—in the one case, the average being 20 above what is regarded as the normal level of prices, and in the other being 11 below. We further observe that the average of prices never again reached the high point of 1820; that even in 1825 it was 3 below, and that after 1825 the average was never very much above what is considered the normal level, except in 1873. After 1832 prices remained below the normal level until 1836. Even then they only went to 102. They dropped rapidly to 94; but in 1839–40 they had risen to 103. But then they fell again, until in 1843 they were as low as 83. A rise once more, till in 1847 they were 95, but fell quickly, until in 1849 the average was as low as 74. Starting, then, with the highest point reached, in 1820 (120), we get in 1849 down to 74—the highest point, as already said, being 20 above what is regarded as the normal level, and the lowest 26 below the normal level. After 1849 there are fluctuations, and then there comes a sharp rise with the Crimean war; a sharp fall in 1857—a year of crisis—a considerable rise after that until 1864, when the average was about 105; a fall in 1870 to 96; and in 1873 a rise to 111. In 1820 the average of prices was 20 above the normal level. In 1825 it was 117, in 1857, 105; in 1864, again 105; in 1873 it was 111. But on the other hand, in 1832 the average was as low as 89; in 1843 was as low as 83; in 1848 was as low as 74; in 1851 was 75; and in 1858 was 91. It will be seen, therefore, that the average of prices was in some cases considerably more under the normal level than it was ever above the normal level, and that the years in which it was under were as numerous as the years in which it was over during the time that French bimetallism existed in its integrity. It is notable, too, that in 1849 the average of prices was as low as 74, and that even last year the average was only 68; so that the average last year was not so greatly under the average of 1849 as is generally supposed. Yet in 1849 French bimetallism was in full vigour. Last year silver was discredited all over the world, had ceased, in fact, to be a standard of value in any civilised country.

Is it not evident from this brief review of Mr. Sauerbeck's chart that the course of prices was not determined by bimetallism; was not determined by the employment of silver as standard money; was not determined even by the quantity of money employed in the world?—that what regulates prices is something altogether different from either the kind of money employed or the quantity of money in existence? I do not deny that money has an influence upon prices. But it is plain from the history of prices from 1820 to 1873 that the influence of money upon prices is much less powerful than the influence of other causes, and that, therefore, it clearly follows that even if silver could be restored to the position it occupied in the monetary systems of the world before the Franco-German War, it would not be a necessary consequence that prices should rise. Prices might rise or they might not, or they might fluctuate violently; but a mere increase in the volume of money would not cause them necessarily to go up, and would not maintain them if they were put up by speculation in anticipation of a general rise. I propose to discuss, next week, some of the causes that act upon prices. But I have thought it useful here to point out that bimetallism while it existed did not maintain high prices, in the hope that those who are clamouring for a change in our currency system without having given long and serious thought to the subject, and in the mere hope that prices may thereby be made to rise, will reconsider what they are doing. Both in the United States and in India we have had lessons as to how tampering with the currency may injure the prosperity of great countries. It is to be hoped that the lessons will not be thrown away upon our own people.

IN the foregoing I have shown from a chart, not certainly drawn up in the interest of monometallism, that the movement in prices since 1820 has been downwards; that between 1820 and 1873 there were very violent fluctuations, but that yet, in the most inflated years, prices never reached the high level of 1820, while in very depressed years they sometimes fell nearly as low as last year's. From this it follows clearly that influences have been at work all through the century tending to depress prices. And it is not difficult to discover what those influences have been — namely, the great mechanical inventions, which have completely changed the economic condition of the world. "Price" is value expressed in the standard money of a country. Therefore, prices must be determined by two sets of causes : by those which affect the values of commodities separately and generally, and those which affect the value of the standard money of the country. But, speaking generally, the former causes are the most powerful. During the present century, at all events, I think no one who has given attention to the subject will dispute that they have been far more powerful than the purchasing power of the standard money of the world. It may be worth while briefly to point out—without any pretension either to exhausting the list or tracing the full effect of these causes—what the principal of them were and how they operated.

In the early part of the century locomotion was extremely difficult, tedious, and costly. It took months to communicate between Europe and the Far East. It took many weeks to communicate between Europe and Eastern America. And such communications were not only slow, tedious, and costly, but they

---

* From *The Statist* of June 9, 1894.

were also uncertain, through the variableness of the winds, as well as dangerous. Therefore it was necessary for merchants to keep large stocks of commodities always in their warehouses. It took so long a time to obtain fresh supplies that it was requisite to keep stocks that would last many months—sometimes even a year or more. Under these conditions prices were necessarily high. Transport by land was exceedingly costly ; by sea it was less costly, but it was tedious, long, and uncertain, and, therefore, the cost of production was increased by the extremely heavy cost of transport. When goods reached here they had to be stored in great warehouses by wealthy merchants, and the merchants necessarily required a profit upon their large outlay and lock up of capital in thus keeping large stocks of goods always on hand. So great a capital, in fact, was required by the merchants that there was comparatively little competition, and the great merchants had matters very much their own way. The introduction of the railway, the steamship, and the telegraph, with such other changes as the opening of the Suez Canal, have completely altered all this. As has been often pointed out in commenting in *The Statist* upon the constitution of the Bank of England, the great mercantile houses are dying out. The functions performed by them are becoming unnecessary. Everyone can now compete who has even moderate capital or moderate credit. There is no need for holding large stocks. A merchant can telegraph to any part of the world for any supplies he requires in a few minutes. In a week or so he can have his goods from America; in a very few weeks he can have them from India. Practically, New York is now nearer to us than Dublin was at the end of last century, and Bombay is far nearer than Lisbon was. The great merchant, then, is rapidly disappearing. There is now no need for the vast capital he formerly employed. Competition is as free in the foreign as in the home trade, and the cost of transport has been immensely reduced ; while the dangers of transport have decreased in proportion, uncertainty having very nearly ceased.

The effect upon trade of the shortening and cheapening of

locomotion has been greatly heightened by the opening up of China and Japan, itself largely the result of the new inventions which have changed the face of the world. And the effect upon trade has been still more heightened by the growth of new countries. In the early part of this century the United States consisted of only a narrow strip along the Atlantic Coast of North America; now the States stretch from the Atlantic to the Pacific. Every part of the country is served by railways, and though the settlement is sparse, yet most of the Continent is settled after a fashion. At the beginning of the century the colonising of Australasia had only just begun. Now Australasia has as large a population as the United States had when the war of 1812 broke out. Canada has grown almost as quickly as Australasia, and so have our South African possessions. The growth of South America has not been quite so satisfactory, but it has been very remarkable for all that. The development of the new countries was rendered possible only by the steamship and the railway. Practically the great emigration from Europe did not begin until the Irish famine of 1847. It received a new impetus from the revolutions of 1848 upon the Continent, and it has gone on upon an immense scale ever since. Some of the most enterprising of the European populations have thus been transferred to the newer countries of the world, have extended cultivation in a way never dreamt of before, and by their competition with the old countries have necessarily increased supplies of produce of all kinds in an unprecedented way, and so have reduced prices.

While all this was going on, the inventions out of which it arose were being improved and supplemented. Labour-saving machinery helped to multiply manyfold the industrial energy of the European settlers in the newer countries, and the perfection of machinery in Europe enabled less labour to turn out far more manufactured goods than before. Every year during the century some advance in this way has been made ; and as these successive advances are being added to one another, the command that mankind has now over the forces of nature is out of all proportion greater than it

LIBRARY
OF THE
UNIVERSITY
OF CALIFORNIA

has ever been before in the history of the world.   This in its turn
has necessarily reduced the cost of production.   And in the meantime the great development of banking has stepped in to render
additional service in the increase of production and the cheapening
of prices.   Every civilised country is now covered with a network of
banks, whose business it is to gather in small sums from depositors
and to lend those sums out again in the great industrial centres to
producers and transporters who may not have got enough themselves.   The banks, it is quite true, are a mere conduit pipe
between the saving classes and the producing classes.   But the
function they perform is not the less valuable or the less powerful
for that.   Without them the business of the world could not now
be performed : with them the more thrifty classes have a new
incentive for saving, not hoarding as their grandfathers used to
do, and their savings are placed by the banks at the disposal of
everyone who is able to add anything to the wealth of the world.
Credit, in fact, has so supplemented capital that it is now the more
important and the more fruitful instrument of the two.   And,
wisely used, credit necessarily tends to reduce prices, not only by
enabling new classes to compete, but by placing the obligation
upon those who get credit to turn over what they deal with rapidly
and frequently.   The origin of the economic revolution effected
during the century is, I need not repeat, the new inventions ; and
of those inventions the most fruitful, and the most powerful by
far have been the railway, the steamship, and the telegraph,
enabling the more enterprising of the European populations
to transfer themselves and their energies from the older and
the slower countries to the newer, to open up those newer
countries, to extend cultivation, and finally bringing the newer
countries in point of time so near to the old that they have been
able to compete successfully with the older, and thus to reduce
extraordinarily the prices of all kinds of raw produce.   But the influence of labour-saving machinery must not be overlooked, nor
that either of banking development.
   The change in prices has been brought about mainly by lower-

ing the cost of production. It is quite true that the cost of production does not determine values. But a lowering of the cost of production enables the new producers to compete more actively with the old ; and keen competition always results in a lowering of prices. Thus, though the cost of production has directly nothing to do with value, indirectly it has an enormous influence upon it, inasmuch as it either stimulates or cripples competition. From this brief sketch it will be seen that the whole tendency of modern civilisation is to lower the cost of production, thereby to stimulate competition, and as a consequence to reduce prices. This is leaving out of account altogether the value of the standard money in which prices are reckoned. But I may remark in closing that the effect of the new inventions did not make itself fully felt until after the Franco-German War. Railway building on a great scale only began here at home in 1843-4. It did not begin upon the Continent until the Fifties. It was later still in the United States and in the newer countries. Roughly, it may be said that the construction of the great main lines continued from about 1843 to about 1873, and that it was not until after the latter year that the full effect of railway construction was felt by trade. Further, it is to be recollected that the Suez Canal was not opened until 1869. And lastly, it is to be borne in mind that the application of steam to navigation and the great changes in marine construction are more recent even than the railway. Some of the changes in marine construction are quite new. The laying of the telegraph, too, is recent. It was not until 1866 that Europe and America were connected by telegraph. All that being so, it necessarily follows that the fall in prices which was going on from 1820 onwards received a new impetus in the Sixties and Seventies, and has become much more marked since. It is these causes and not the breakdown of bimetallism which have occasioned the great fall of the past twenty years.

THE APPRECIATION OF GOLD.

GOLD Monometallists contend that the new inventions which have characterised the closing century fully account for the fall in prices, and that when so simple and sufficient a cause can be adduced, it is unscientific to bring in a more complex and more doubtful cause. To that I reply that the cause is certainly not simple, and that its action is not uniform. The prices with which we are all dealing in this controversy are the wholesale prices of commodities, and the new inventions of the century have affected commodities in two ways. As I have endeavoured to show in the foregoing, they enable capital to go farther than it formerly did in producing, transporting, and distributing. To that extent it is perfectly clear that the new inventions have lowered prices. The money outlay being smaller, competition may be trusted to ensure that the money return shall be smaller likewise. Secondly, the new inventions have increased the efficiency of labour; but if the efficiency of labour has been increased as much in gold mining as in the production of other commodities, then it is evident that the mere increase in the efficiency of labour cannot have changed prices at all. If the standard money of a country represented labour, it is unquestionable that the more efficient labour became the more would the standard money be appreciated. But the standard money of no country represents labour, and whether, therefore, the standard money of a country is or is not appreciated by the increased efficiency of labour, depends upon whether the efficiency has or has not been increased in the same proportion in gold mining as in other industries. But it will hardly be dis-

* From *The Statist* of June 16, 1894.

puted that labour is not as efficient, or at all events has not been as efficient, during the past quarter of a century or so in gold mining as in other industries.

The mistake both of the gold Monometallists and of the Bimetallists seems to me to spring out of their adherence to the quantitative theory of money, which I venture to think is not merely misleading, but is utterly wrong. Save in a few instances in which the quantity of a thing cannot be increased, such, for example, as paintings by the old masters, or where there is a real monopoly, the value of a commodity is not determined by the existing quantity, but by the cost of that portion of the new supply which is most expensive to raise. Suppose, for instance, that the people of the United Kingdom insisted upon having every year 28 million quarters of wheat, and that 27 million quarters could be produced and sold at 20s. per quarter, but that the last million quarters could not be sold at less than 24s. Then the price of wheat must be 24s. a quarter or somewhat more. We all know, of course, that the difficulties of producers and intermediaries, mistakes for a while as to supply and demand, accident and the like, may cause great fluctuations in prices. The price of wheat might fall for a while considerably under 24s., or might rise considerably over that figure, but it is certain that if the people of the United Kingdom insisted upon having the full 28 million quarters every year they would have to pay the price that would induce the growers of that portion which is produced at the greatest cost to continue cultivating. So again let us suppose that a manufacturer in one of the great manufacturing centres is employing at present a thousand workpeople; that suddenly there is an increased demand for his goods, and that he requires fifty workmen more. The wages he will have to pay will not be determined by what he is already paying the thousand men, but by what will attract to him the fifty men whom he suddenly requires. And so it is with every commodity. But gold is a commodity just as iron and tin, and the value of gold is determined in precisely the same way. The return to the gold

miners must be sufficient to induce those who are engaged in raising the gold from the costliest mines to continue working. It is
perfectly true that gold mining is of the nature of a speculation, that
it has a peculiar attraction for certain minds, and that the owners
of mines are very reluctant to cease working because of the extraordinary expense of reopening. It is quite possible, therefore, that
mines may be worked at a loss for a considerable time, but while
fully granting that, it is equally certain that men will not go on
working at a loss for a quarter of a century or more, and that
therefore they must get the usual return to labour and capital.

The Bimetallists reply to this, that while the theory is true
enough in regard to ordinary commodities, it is not true of gold
and silver, because of the enormous quantity of these two metals
existing in the world, and their extraordinary durability. Let us
then put the reasoning in a somewhat different way. We are
dealing, it will always be recollected, with the wholesale prices of
commodities, and in the wholesale markets it is notorious that
metallic money is hardly employed. A man goes to his banker
and raises a loan upon security. In 999 cases out of a thousand
he does not draw cash. The amount is credited to him in the
books of the bank. He buys commodities and rarely pays cash.
Probably he pays by means of a cheque. And the seller also
usually does not take cash. The price is credited to him either
in the books of the same bank or in the books of some other
bank. Similarly goods are bought by means of bills, and the bills
are settled through the Clearing House without the intervention of
cash, except to an infinitesimally small amount. It is unnecessary to extend the argument or to quote evidence. Everybody
knows that in this country metallic money is used very little—to
quite a small fractional amount in the wholesale markets, and
even on the Continent to a very small extent. But if the prices
of commodities are paid, not in cash, but in credit documents, it
is perfectly clear that the quantity of money cannot determine
those prices. They are clearly determined immediately by the
volume of credit, and ultimately they are governed by the value

of the standard money in which they are reckoned, although not discharged. I submit, then, that whether we look at gold as a commodity pure and simple or as the substance of our standard money, it is not the existing quantity of it which determines its value, but the cost of raising the most expensive portion of it which is required at the time.

A certain annual supply of gold is always required, partly because of the growth of wealth and population, partly because of the loss through wear and tear and accidents, and partly because of the consumption of the metal for the purposes of the arts. During the past four-and-twenty years the amount of gold required has been exceptionally great, owing to the adoption of the gold standard by Germany and so many other countries. On the other hand, the production of gold fell off immensely immediately after the Franco-German war. It had been declining for some years previously. It decreased all through the Seventies, and, according to Herr Soetbeer, during 1881-86 it averaged less than £20,000,000 per annum. There having been, then, in addition to the normal demand, an extraordinary demand for supplying new currency for the countries that adopted the gold standard, for hoarding, for war chests, and for India, it follows that the value of gold must have risen. No serious man would dispute that if the consumption of any other commodity had immensely increased during twenty years, and the supply had fallen off, there would have been a marked rise in the value. And I have just been endeavouring to prove that gold conforms in every respect to the laws that govern the value of all other commodities. In these days of keen competition and of close communication between all parts of the world, is it conceivable that immense amounts of capital and labour would be employed for a quarter of a century or so at a return less than could be obtained from any other industry? It is undisputed that wages, generally speaking, have not fallen in the great industries of the world since the Franco-German war; but unless gold increased in value wages must have fallen in gold mining, for whereas in almost all other industries the return to the labourer—that is to say, the return per man in production—has enormously increased, the return per man in production in gold has unquestionably greatly decreased. If, therefore, the labourer in a gold mine was to get as good a wage as the labourer in other industries, he must either have taken so large a proportion of the gold outturned as to leave nothing to the capitalist, or a smaller gold return must have given him an increased command over commodities generally.

THE THEORY OF PRICES.*—III.

I HAVE endeavoured to show above that gold has appreciated, and that its value is not determined by the quantity of the metal already in existence, but by the cost of raising the most expensive part of the required supply; that prices are determined immediately by credit, and are regulated ultimately by the value of the standard money of the country.   It may be asked, " How does the value of gold affect credit ? "   The answer is simple enough. The volume of credit at any time is determined by two sets of causes—those which affect the persons who demand credit and those which affect the persons who grant credit.   Of these causes, by far the most powerful is opinion—the temporary opinion as to the solvency of people generally who are engaged in business, or what is usually spoken of as the state of confidence in trade. Unless there is confidence very few people will care to engage in new enterprises or to extend old businesses.   There will therefore be very little demand for credit.   And, similarly, bankers, who are the great grantors of credit, will not care to give very much.

---

* From *The Statist* of June 23, 1894.

E

Confidence, then, or more generally and broadly speaking, opinion, is the most powerful of all causes influencing credit. But in any state of credit, whether good or bad, it is clear that the amount of the bank reserves must have an influence upon it. In countries like our own it is evident that, however good credit may be, a banker can only lend in proportion to his resources. Leaving out of account banks that are managed with criminal recklessness, all bankers must and do keep some kind of reserve. The proportion differs from country to country, and even in the same country there is much difference of practice in the keeping of reserves. But every respectable banker keeps *some* reserve ; and it is clear, therefore, that there must be a limit to the amount of credit he can give. As soon as it seems to him that his reserve is becoming dangerously small he must stop adding to the magnitude of his loans and discounts. Consequently, the amount of the reserves limits the amount of credit that can be granted, and, therefore, when the standard money forming the reserves is scarce—or, rather, when the new supplies of the metal forming the standard money are smaller than usual—reserves tend to become small and so to limit the volume of credit.

But it is often objected by gold Monometallists that if this were so the rates of interest and discount would have been exceptionally high during the past quarter of a century. Not at all. I venture to submit, on the contrary, that when gold is appreciated, rates must be lower than when gold is depreciated ; and this because the smaller the reserves are the sooner they run down to the danger point, and all the sooner, therefore, bankers put a check to speculation by refusing to lend and discount. Everyone knows how, when bankers think that speculation is becoming dangerous, they call in loans. When they do that they may make rates dear for the moment ; but they create a fear amongst speculators which compels the latter to liquidate their accounts and so, for a considerable time afterwards, speculation is checked. To illustrate this point, let us suppose that an extraordinarily rich vein of gold were to be discovered, that the outturn were to be so immense that for a considerable time a million sterling was sent in at regular intervals from this new field to the Bank of England. Instantly the opinion would spread that money was about to become unusually plentiful and cheap, that prices must rise, that business must expand ; and consequently everyone engaged in business would conclude that if he bought goods then, he would be pretty sure to make a profit. Speculation would consequently spring up both upon the Stock Exchange and in the ordinary commercial markets, business would expand immensely ; and if the new supplies of gold continued on an extraordinary scale the banks would be in a position to go on week after week, adding to the accommodation they give their customers. The rise in rates that would follow might last for a considerable time. Rates might reach five or six per cent., or even more, and be maintained for a much longer period than we have been accustomed to for the past twenty years. Now suppose, on the other hand, that gold were to be taken away from the Bank of England instead of being sent into it, and if it were to be taken in large amounts, bankers would soon become apprehensive, and would begin to call in loans. Every prudent business man would then begin to lessen his commitments, and those who are less prudent would be required to do so by the calling in of loans by bankers. During the past twenty years the gold reserve of the Bank of England, speaking generally, has been exceedingly small in proportion to the immense volume of business done by this country. Consequently the instant we have had a beginning of really active speculation bankers have become apprehensive, loans have been called in, and the rise has been checked. We have never during this period had prices at their old level, because bankers were not in a position to give credit in proportion to the country's volume of business on the same scale as they were before the Franco-German war.

But it may be asked—allowing for the sake of argument that prices are determined by credit, and that the volume of credit is limited by the amount of the reserves held by the banks—how can the value of gold be so transmitted from the reserves to credit as to affect prices? The answer is not difficult to find. Gold is required for four distinct purposes—for the arts, for hoarding, for the general circulation, and for bank reserves. Gold used in the arts tends, of course, to raise the value of the metal; but, except in that way, it does not affect prices. It is not employed as money. Gold hoarded, in the same way, by increasing the demand for the metal, raises its value. But while it is hoarded it is out of the reach of trade just as much as if it were still in the mines, and therefore cannot affect prices. Even the gold employed in the general circulation does not affect prices directly. It is performing the function of small change, or token money just as much as silver coins are. It is employed, it is true, in paying wages, and it is also largely used in the retail trade. But, speaking generally and roughly, it may be said that it is not used in the wholesale trade. The only gold that affects prices directly and immediately is that in the bank reserves. If the supply of gold is large enough to meet the demand for the arts, for hoarding, and for the general circulation, and yet to leave a liberal amount for bank reserves, the reserves will be large, and consequently the superstructure of credit built upon them will be large. If the reserves be small, so will be the superstructure of credit built upon them. Further, in the latter case gold will be eagerly sought for everywhere, and the most expensive mines will be carefully worked. The value of gold will therefore almost inevitably be high. But prices are measured in gold although they are discharged, generally speaking, by means of credit documents ; and although it is very rarely that gold is actually used in making payments, still every business man who has a right to receive gold may demand it in payment. Therefore prices may not be higher, even though they are fixed by credit, than if they were necessarily payable in gold. In other words, the credit prices will be measured in gold and will be regulated by the value of gold.

LORD FARRER'S LETTER.†

1 AM not sure that I rightly understand Lord Farrer's letter in last Saturday's *Statist.* He is so kindly considerate that he did not give himself room to fully explain his meaning. But if I do understand him, I submit that his admissions dispose of his contention. He writes : "The nexus, and the only nexus, between an increase or a diminution in the quantity of gold for the time being available for currency on the one hand, and wholesale prices on the other, is to be found in the operation of that increase or diminution on the state of the bank reserves." I know of only one way in which the scarcity of gold can affect wholesale prices through the bank reserves. It is, as I explained a fortnight ago, firstly by limiting the bank reserves, secondly by limiting credit thereby, and thirdly by lowering prices in consequence. But the rates of interest and discount are simply the rent paid for the use of loanable capital ; and the rent of loanable capital, like the rent of land or of anything else, cannot be high when there is such a scarcity of gold as makes all prices low. Speculation may be likened to a railway train, which travels at a great pace when stopped only at long intervals, and crawls along when stopped very frequently. Speculation must be pulled up frequently when the scarcity of gold so contracts the bank reserves as to make the superstructure of credit raised upon them small. I venture to think that Lord Farrer's mistake, as well of that of the gold Monometallists generally, arises out of the use of the ambiguous word "money," instead of "loanable capital." Through that

---

* From *The Statist* of July 7, 1894.
† See Appendix A.

error he is led to confound loanable capital with cash, when in reality loanable capital is neither more nor less than credit.

In the last paragraph of his letter Lord Farrer writes : " If therefore, during the period of low prices which has continued during the last twenty years or more, we find that the supply of gold in the bank reserves has been as abundant as ever, and that the rates of discount have not been higher, and have not fluctuated more than they did in previous years ; and if it cannot be shown that the lowering of prices has been preceded or accompanied by a diminution in the gold reserves or a raising of the rate of discount, a strong presumption—to say the least—is raised that the low prices have not been due to an increased scarcity of gold." This extract clearly discloses the ultimate cause of Lord Farrer's error. During the past twenty years the volume of trade of the whole world has increased immensely compared with the preceding twenty years. How much the increase has been I admit is very difficult to calculate. But that there has been a very great expansion of trade, I presume no competent person will deny. Unfortunately, the only statistics available to help us are those relating to the international trade of the world. I submit, however, that the international trade has not grown at all as rapidly as the home trade of the more advanced countries. Our own country has the greatest foreign trade of any nation in the world ; and yet I do not hesitate to express my own opinion that our home trade has expanded during the past twenty years more than our foreign trade ; and I have no doubt at all that the home trade of such countries as the United States, France, and Russia has grown far more rapidly than their foreign trade. But, even if we take the international trade of the world alone, it is incontestible that the volume of the world's trade has expanded greatly during the past twenty years. If Lord Farrer is prepared to admit that, is he prepared to dispute that larger gold reserves are required now to maintain the level of prices at the level of the twenty years ended with 1873 ? Is he prepared to deny, further, that the bank reserves have not increased during the last twenty years in anything like the proportion in which the world's trade has expanded ? If he admits the first proposition and cannot deny the two last, does it not follow that the failure of the bank reserves to increase —at least in the same proportion as the volume of the world's trade—must necessarily have led to a contraction of credit, and so to a fall in prices ?

During the past twenty years silver has been demonetised by country after country, so that practically the civilised world has come to depend upon gold alone for its bank reserves. Further-more, the new supplies of gold from the mines had fallen off greatly until three or four years ago. And, lastly, the consumption of gold has immensely increased. Is it not reasonable to assume that since silver has been discarded by so many countries as standard money, since gold has been consumed in such increased proportions, and since the new outturn of gold from the mines had been falling off seriously for so long a time, it would not be enough to have the same proportion of bank reserves to the volume of trade which was thought to be sufficient in the preceding twenty years ? Personally I have no doubt that the so-called scramble for gold, which has exercised so great an influence upon the opinion of the trading community all over the world, has made it necessary to keep much larger reserves, and that, therefore, to maintain the level of prices at the average of the twenty years ended with 1873 we ought to have a much higher proportion be-tween the bank reserves and the volume of trade than was sup-posed necessary formerly. But I do not wish to push the point too far, and I am willing, therefore, for the sake of argument, to assume that if the bank reserves had grown in the same propor-tion as the volume of the world's trade, both national and inter-national, the level of prices would have been maintained. But it is notorious that the bank reserves have not grown at the same rate as the world's trade ; and, consequently, I submit that a decline in the level of prices was inevitable.

There is one other point to which I would like to invite Lord Farrer's attention, and it is the difficulty of ascertaining what the bank reserves are at the present time. If he will look at *The Statist's* tabular Appendix he will see that the Imperial Bank of Russia holds over 56 millions sterling in gold, besides a mixed sum of gold and silver. I presume he will grant that these 56 millions sterling are not a "banking reserve" in the sense in which both he and I employ the phrase. They are a hoard or a war treasure ; but they are not a "reserve," kept in ordinary times to meet trade contingencies. So, again, the gold held by the Austro-Hungarian Bank is not a reserve in the proper sense of the term. When resumption is completed it may become so ; but at the present time it also is a hoard, not a reserve. Now let us come to the Bank of France. It holds over 72 millions sterling in gold, and over 51 millions sterling in silver—together over 123 millions sterling. But I presume Lord Farrer will concede that the silver forms no part of the Bank's reserve. It is coin locked up in the Bank of France for the purpose of keeping the legal tender five-franc pieces at their legal value. How much of the gold is a reserve ? How much must be kept by the Bank of ·France in order to assure the peoples of the Latin Union that the 160 millions sterling of legal tender silver coin will be maintained at the mint value ? How much, again, must be set aside as a war treasure ? The French Government, unlike the German, does not hold gold in any of its fortresses—or, at all events, it does not let the world know that it holds a special war treasure. The real war treasure is kept by the Bank of France. Suppose the great war which Europe has been dreading for so long a time were to break out, the expenditure would exceed anything the world has ever yet known. Would it not be absolutely necessary that the French Government should be able to lay its hand without an hour's delay upon a vast sum which would be received everywhere abroad as well as at home in payment for what it required ? Otherwise the French Government would have to borrow on an unprecedented scale. And if Russia were the ally of France is it not certain that France would have to provide unheard of sums for Russia also ? How much, then, of the gold held in the Bank of

France must be ear-marked as war treasure which the Bank cannot use in any way without the permission of the Government? And when we have made allowance for the war treasure and the reserve against silver depreciation, how much gold remains as a real banking reserve?

One last point. What is the reserve of such countries as Egypt, for example? Is it not held by the Bank of England just as much as the reserves of the Scotch and Irish banks? Again, what is the reserve of such countries as Argentina? Is it also not held by the Bank of England? Is not, then, the reserve of the Bank of England, which looks so immense to the unthinking just now, a reserve not only for the United Kingdom, but for a very large part of the outside world as well? If Lord Farrer will add up the gold held by the reserve-keeping banks of the world and will then make allowance for war treasures, for depreciated silver, and for the requirements of foreign countries which have no reserves of their own, I venture to think he will find that the bank reserves at the present time are not merely small but are *very* small; and that when he takes into account, on the other side, the vast growth in the world's trade and population, the burden of debt that is hanging like a mill-stone round the necks of so many states, and the fear of the most terrible war that the world has ever seen, he will recognise that a fall in prices was inevitable from the depletion of the bank reserves alone.

THE THEORY OF PRICES.*—V.

LORD FARRER'S SECOND LETTER.†

LORD FARRER, I hope, will pardon me if I say that his second letter is disappointing. It purports to be an answer to the articles I have been contributing to *The Statist* on the bimetallist agitation, and more particularly to the article published a fortnight ago. I appreciate the compliment, and am anxious to conduct the controversy with fitting courtesy ; but I am bound in truthfulness to say that Lord Farrer has not taken the trouble to understand the argument he undertakes to refute. For instance, he writes last Saturday : " Your own hypothesis is that the way, and the only way, in which a diminution in the quantity of gold available for currency can affect prices is through the rate of discount." I hold the very contrary, and I have already written two articles to disprove the proposition which Lord Farrer puts into my mouth. My point is that the rates of interest and discount cannot be high permanently while gold is scarce and dear, any more than the rent of land or the prices of commodities, for, in fact, interest and discount are the rent paid for the use of loanable capital. I challenge Lord Farrer to show *why* the rent or consideration paid for the use of loanable capital should be high, except very temporarily, when the rent of land and the prices of commodities are low. I go farther, and challenge him to show *how* rates can be high when prices are low because of the scarcity of gold. The truth is that the gold Monometallists, by using the ambiguous and misleading word " money " when they mean loanable capital, have confused themselves, and have jumped to an absolutely untenable conclusion.

---

* From *The Statist* of July 21, 1894.
† See Appendix B.

Lord Farrer has taken offence because, very innocently, I said what I have just repeated; that the gold Monometallists have confused themselves by using an ambiguous and misleading term. I certainly had no intention to give offence, and it is far from my wish to be uncourteous. But if Lord Farrer will allow me to say so without taking umbrage I submit that his last Saturday's letter proves that the misleading word has confused him. He passes in a bewildering way from gold as a medium of exchange to gold as the substance forming the reserves of the great reserve-holding banks of the world—two entirely different things which Lord Farrer could not possibly confound, if he would drop the use of the word "money" and employ instead either "loanable capital" or "credit." Gold as a medium of exchange is simply part of the currency of the world, and while it is circulating as such it cannot also form part of the reserves. The currency of any country may consist of either gold or silver or copper, or paper, or of all four; but the gold, while it is in circulation, is performing only the same function as token coins or credit documents. It is, no doubt, a potential reserve. It may be withdrawn from the circulation and locked up in the reserve-keeping banks; but while it is in circulation it is no part of the reserves. And when gold is largely used as currency less of it is left for the bank reserves. Surely Lord Farrer is aware that besides its use in the arts, gold is employed in four distinct ways—as a circulating medium, as a war treasure, as a hoard, and as a banking reserve. And the fact that it is so employed does not prove that gold is plentiful; on the contrary, when the new outturn from the mines is small it goes far to prove that gold is scarce. The larger the hoards, the war treasures, and the circulating media, the smaller are the supplies available for banking reserves. But I contend that it is the banking reserves alone which affect prices, and that the banking reserves have been so attenuated during the past twenty years that the fall in prices was a necessary consequence. Another instance of the confusion of terms is afforded by the paragraph in Lord Farrer's last letter where he argues: "The changes in the modes of doing business and the growth of credit, make it possible, and even necessary, for an ever-increasing amount of business to be done safely and properly upon the basis of an ever-diminishing proportion of gold." Here he is evidently thinking of gold employed in making payments. It is perfectly true that in the progressive countries gold is employed less and less in making payments. It is hardly employed at all in the wholesale trade and upon the Stock Exchange, and it is becoming less and less employed even in paying salaries and in the retail trade. Cheques and other credit instruments are taking its place. But the greater the development of credit, the more important is it that the bank reserves should be large. I dissent altogether, therefore, from Lord Farrer's proposition, and maintain that, if prices are not to fall, every growth of credit requires a corresponding growth of the bank reserves. I grant, indeed, that our own banks have been exceedingly remiss in the keeping of reserves. The Bank Charter Act is barely fifty years old, yet three times it has had to be suspended, and the need for a fourth suspension would have arisen in 1890 were it not that the Bank of England went begging for help to the Bank of France and the Imperial Bank of Russia. That certainly does not prove the necessity for, or the safety of, small reserves.

Lord Farrer cites at considerable length Dr. Soetbeer and Dr. Neuman Spallart to buttress his assertion that gold has not been scarce since 1873. Both authors were very careful in the collection of facts, and I am quite ready to accept their figures as figures only. But the inferences they draw are simply the inferences of intelligent men, and are worth no more than the conclusions drawn by any other statistical students. In this particular case I believe they are both utterly wrong, and in the article to which Lord Farrer more particularly is replying I gave my reasons for the opinion. Lord Farrer does not think it necessary to touch upon those reasons, and yet I submit that no man can understand the economic condition of the world during the past quarter of a century who does not carefully discriminate between the gold held as banking reserves, and the gold held as war treasures. For example, the Government of Russia has at the present time nearly 100 millions sterling in gold, and yet that is not a gold reserve, and there is no intention to use it as such ; at all events, for many a year to come. Again, the Austrian and Hungarian Governments have been withdrawing large sums of gold, chiefly from the United States, during the past couple of years. In the United States the gold was held by the Treasury and by the banks, and was a banking reserve. In Austria and Hungary it is not a bank reserve; it has been completely withdrawn from banking purposes for the time being. Of course, if war does not break out, and if Austria and Hungary carry through their monetary reform, the gold will become a banking reserve. But at the present moment it is not such ; it is either a war treasure or a hoard accumulated in preparation for the resumption of specie payments, whichever way you may choose to look at the facts and interpret the policy of the Austrian and Hungarian Governments. The banking reserves proper are very small. What is really peculiar about the monetary history of the past twenty years or so is not the accumulation of vast bank reserves, for there are no vast bank reserves anywhere ; but the accumulation of vast war treasures by so many of the leading countries of the world in anticipation of the struggle which they have been preparing for in every way throughout the whole time. Just as they have been piling on debt and taxes, and bringing under the cane of the drill sergeant larger and ever larger proportions of the population capable of bearing arms, so they have been withdrawing from the service of trade more and more of the gold that would otherwise have gone into the bank reserves.

THOSE who agree in the main with the argument laid down in the foregoing articles will see now that in its very nature bimetallism is impracticable. The Bimetallists deny this, grounding the denial on the existence of bimetallism in France for about seventy years within the present century. But, as I have shown, the ratio of 15½ to 1 was never maintained during the seventy years; one or other of the metals was always at a premium. And further, France was practically a silver-using country until some time after the great gold discoveries in California and Australia, when she became a gold-using country. But the bimetallists reply that it is not pretended that a ratio can be strictly maintained; all they assert is that wide fluctuations can be prevented. It is true that the values of gold and silver did not fluctuate very widely up to about 1873; but that was due not to French bimetallism, but to the fact, firstly, that up to 1834 all the world was practically silver-using, with the exception of the United Kingdom; that from 1834 to the middle of the century all the world was practically silver-using except the United Kingdom and the United States; and that when gold began to be produced in Russia, in the United States, and in Australia in vast quantities, about the middle of the century, France became gold-using and the Far East absorbed the silver discarded by France. A conclusive proof that bimetallism was quite ineffectual is afforded by the fact that the instant Germany demonetised and proceeded to sell her silver, French bimetallism broke down. Thus the history of bimetallism in France and in the Latin Union generally, confirms in the most striking way the experience of the whole world from the earliest times that it is impossible to maintain a fixed ratio

---

* From *The Statist* of June 30, 1894.

between the values of any two commodities, let them be what they may.

That the Bimetallists themselves are not very confident of the practicability of their panacea may be inferred from their inability to agree upon a ratio. M. Cernuschi proclaims that the old French ratio of 15½ to 1 must be maintained. But the English Bimetallists disavow M. Cernuschi, and, in fact, allege that he is a quack. It might be sufficient to say that until Bimetallists can agree amongst themselves upon a ratio, the rest of the world need not trouble with the matter. But I would like to ask in addition whether any Bimetallist has seriously thought out how a fixed ratio is to be established and maintained? Clearly there are only three ways: either the value of silver must be artificially raised, or the value of gold must be artificially lowered, or both processes must be combined. But I would ask: How is the value of silver to be artificially raised? The Bimetallists airily tell us that if the mints of the great nations of the world are opened to the free coinage of silver, the consumption of the metal will so increase that its value will rise, and consequently the price. That is a mere assertion which gives literally no information. The Sherman Bill was introduced into Congress in the spring of 1890, was passed in the middle of July in that year, and came into operation on the 13th of the following month. While the discussion was going on a great speculation sprang up, and the price of silver rose from about 44d. to 54⅝d. per ounce; the highest quotation being reached in the first week of September— just three weeks after the Act came into force. From that moment the price began to fall, and steadily declined until, a little before the closing of the Indian mints last year, the price was about 38d. During this period of nearly three years the United States bought 54 million ounces of silver annually; India imported about 7 millions sterling worth of the metal per annum, and the Mexican mints were as active as they could be turning out dollars for the supply of other parts of the Far East. Yet with this enormous demand the price of silver fell from 54⅝d. to about 38d. Is it not perfectly certain that, even if bimetallism were adopted, the consumption of silver would not be greater for a considerable time after the new system was introduced than it was during the three years the Sherman Act was in force?

It is evident to every careful observer that the currency of the United States is at present redundant. Therefore, either some of the paper money must be cancelled, or some of the gold must be parted with, or some of the silver must be got rid of. At the present time gold is being driven out very rapidly. If bimetallism were generally adopted, and there were to be a free market for silver, we may reasonably conclude that the Americans would keep their gold and get rid of as much silver as they could. Therefore, for the early years of bimetallism America would be a seller and not a buyer of silver. Furthermore, it is not probable that with bimetallism India or the other countries of the Far East would absorb more silver than they did during the three years immediately preceding the closing of the Indian mints ; and it is certain that Mexico could not absorb more silver. But Russia, Italy, Spain, Portugal, and Greece have suspended specie payments and would not open their mints. Neither would the South American countries generally. Consequently, the only fresh demand there could be for silver under bimetallism, would be in the countries which at present are gold-using. Does anyone seriously believe that the gold-using countries would get rid of their gold and absorb the silver that America would be trying to sell, and possibly France also ? There is another point. We have seen that, with the enormous demand for silver caused by the Sherman Act, the price fell to very nearly 3s. an ounce. Is it not probable, therefore, that the whole demand of the world could be supplied at 3s., or, at all events, at 3s. 3d. per ounce ? Indeed, if the Director of the United States Mint is right, silver could be produced much more cheaply, for he tells us that in 1893 the outturn was 161 million ounces, against only 153 million ounces the year before. But if the real value of silver is 3s. or 3s. 3d. per ounce, with the greatest conceivable demand from the mints, how, I would ask once more, do the Bimetallists propose to raise the price to 5s., or 4s., or, let us say, 3s. 6d. an ounce? I would ask any business man of experience. Is it possible by any artificial arrangements of Governments to enable one small class of capitalists and labourers to earn exceptionally high profits and wages permanently? Is it not as certain as anything can be that competition would force the price down to nearly the intrinsic value? And if it did, how do the Bimetallists propose to raise it artificially?

BIMETALLIC FALLACIES.*—II.

A FORTNIGHT ago I asked how the Bimetallists propose to raise the value of silver, supposing all the great nations of the world could be induced to adopt bimetallism; and I brought forward some reasons to show that it would be impossible to put up the value. That there would not be the slightest chance of getting the countries which are interested in silver, and more particularly France, to join in a universal Bimetallic League unless silver was advanced in value seems unquestionable. The Bank of France holds over 51 millions sterling in silver at the mint price of the metal; that is, on the old ratio of 15½ to 1. If bimetallism were now adopted, and silver was rated at its present market value, it is clear that the Bank of France would lose more than half this whole sum—more, than is to say, than 26 millions sterling. The effect, therefore, of adopting the present value of silver by the French Government would be to make the Bank of France hopelessly insolvent. It may be said that the loss would be assumed by the nation. But what equivalent could be offered to France to induce her to incur on account of the Bank of France alone a loss of

---

* From *The Statist* of July 14, 1894.

F

over 26 millions sterling—especially as the loss on the silver outside the Bank of France would be at least as great?

It is generally estimated that the legal tender silver in the nations forming the Latin Union amounts to about four milliards of francs, or 160 millions sterling, at the ratio of $15\frac{1}{2}$ to 1. It does not matter very much whether this estimate is too high by 10 or even 20 millions sterling; it is certain that the amount of legal tender silver in the Latin Union is enormous, and the result of adopting the present market ratio would be to reduce this legal tender silver in value by more than one-half. Is it likely that Italy in her present embarrassment would voluntarily incur the loss of one-half of all the legal tender silver coined by her? France would, of course, be better able to face the difficulty; but what equivalent could be offered to France to induce her to submit to so enormous a loss? It seems, therefore, that M. Cernuschi is perfectly right when he says France never will enter into a Bimetallic League unless the old ratio of $15\frac{1}{2}$ to 1 is adopted. And the Bimetallists themselves admit that if France or any other great nation were to hold aloof from the League, bimetallism would fail. That being so, I ask again, by what process could the value of silver be raised to double what it is at present, or indeed could be raised at all appreciably?

The only answer to this question by the Bimetallists that I have ever seen is that if all the great nations of the world opened their mints to the free coinage of silver an almost limitless market for the metal would thereby be created, and it would automatically rise in value. It is almost inconceivable how intelligent men can be so greatly misled by playing with words. The opening of the mints to any metal, no matter what, does not create a market. The mint simply receives from the holders of the metal bullion offered to it, and places a stamp upon that bullion when coined to testify that it is of the regulation weight and fineness. The mint does not itself retain the coins; it returns them to the person who sent in the bullion, or his assignee, and it is the owner of the bullion who has to dispose of the coins. It would be just as reasonable to say that the tailors who work for an East End sweater buy the cloth which they make into coats and waistcoats for the sweater as to assert that the mint in putting its stamp upon either silver or gold buys the metal from the owner. It is quite true, of course, that the gold or silver while in the form of bullion is not legal tender, and that it becomes legal tender when it is coined. It is unquestionably in a more available shape for commercial transactions when it is coined. For all that, the mere putting it in a more available shape does not ensure that it will be freely accepted. Every manufacturer knows that the goods he makes up for sale very often cannot be got rid of. They are in an available shape for consumption, but for one reason or other a demand for them does not exist : they are left upon his hands, and sometimes the consequent lock-up of capital is so great that the manufacturer is ruined. And what happens in the case of manufactured goods may equally well happen in the case of manufactured coins.

This is shown conclusively by the experience of the United States Government from 1878 to 1890. In the former of those years the Bland Act was passed, requiring the Secretary of the Treasury to buy every month not less than two million dollars worth of silver, and to coin it without delay into legal tender pieces. Everything the Treasury could do was tried to force the silver pieces into circulation, but without effect. After a time the effort was given up, and silver certificates were issued instead of dollars. So complete was the failure to get out the silver coins that in 1890, when the Sherman Act was passed, the experiment was given up. In the summer of 1892 the United States Senate asked the Director of the Mint what was the number of standard silver dollars then in circulation, and the answer was, 56,779,484. The Senate further asked what was the number of standard silver dollars then in the Treasury. The answer was, 357,189,251. Thus, in the fourteen years immediately following the passage of the Bland Act barely one-seventh of the legal tender dollars coined under that Act had been got into circulation, in spite of all the efforts of the American Government. If the American Government so hopelessly failed—with all its prestige and authority, in a country where silver is a great native production and where there is a strong desire to protect native industry—what chance is there in any other country that private persons and trading companies could force silver into circulation even if bimetallism were universally adopted? It may be replied that certificates or notes could be issued against the silver, as they were in the United States. They were issued in the United States, however, by the Government itself; and the proposal of the Bimetallists is not that the Governments of the world should buy and coin silver and issue certificates against it, but that they should throw open their mints to the free coinage of the metal. Private holders of legal tender silver pieces, whether they were individuals or companies, could not issue certificates ; and judging from the experience of the United States Government, they could not themselves get the pieces into circulation.

The chief reason why the United States Government failed to get the silver coin into circulation was the refusal of the banks all over the Union to hold any considerable amount of silver in their reserves. While I was predicting in *The Statist* the crisis that occurred in the United States last year I showed again and again by analysing the returns made by the national banks at various times to the Comptroller of the Currency that as a matter of fact the banks were boycotting silver. Any reader who has a file of *The Statist* by him will see the figures set out in detail in the issues of July 23, 1892, September 17, 1892, and November 26, 1892. In the last of those articles more particularly, I showed that the silver coins and silver certificates together held by all the national banks throughout the Union amounted to only between one-eighth and one-ninth part of the gold and paper redeemable in gold held by the banks. It will be recollected that silver was just as much legal tender as gold in the United States at the time, that gold was being shipped to Europe in very large amounts, that it was being withdrawn from the banks as well as from the

Treasury, and that, therefore, it would have been much easier for the banks to replenish their reserves by means of silver than by means of gold. But, in spite of that, the banks boycotted silver and exerted themselves to the utmost to increase their gold holdings. What happened in the United States would happen in Europe if bimetallism were adopted. The banks would refuse to hold silver in their reserves ; silver coin consequently could not be forced into circulation ; and, that being so, I ask once more how the Bimetallists propose that the value of silver should be raised if universal bimetallism were adopted ?

BIMETALLIC FALLACIES.*—III.

A FORTNIGHT ago I showed reasons for believing that M. Cernuschi is right in saying that France never will enter into a Bimetallic Convention unless the old ratio of $15\frac{1}{2}$ to 1 is adopted ; and I put the question which I had previously asked, how the Bimetallists propose to raise the value of silver supposing —what to me seems incredible—that the nations could be induced to enter into a universal Bimetallic Convention, and to agree to the ratio of $15\frac{1}{2}$ to 1. If the value of gold were to remain as high, or nearly as high, as it is at present, and the value of silver were to be somewhat more than doubled, it is clear that the object the Bimetallists generally have in view would be defeated. Their contention is that there is not enough of gold to supply the world's monetary requirements, that the consequent fall in prices has inflicted ruinous losses and has brought about the present depression. Therefore they propose that silver should be rehabilitated in the hope and belief that prices would

---

* From *The Statist* of July 28, 1894.

thereby be raised. But unless gold were considerably reduced in value, this would not happen. In the Latin Union at present, and in the United States there is a very large amount of silver legal tender money and silver notes ; yet gold is the real standard of value, and prices are regulated by the value of gold just as much as in our own country. If, therefore, the value of gold under a bimetallic arrangement were not greatly reduced, the condition of all the world would be similar to the condition of the Latin Union and the United States. Gold would practically remain the standard of value, and silver would be a subsidiary money no matter what hocus-pocus were applied. By what process, then, I ask once more, do the Bimetallists propose to lower materially the value of gold ? Personally I have no doubt that we are on the eve of a fall in gold, or, what is the same thing, of a rise in general prices. Distrust, occasioned by the unsatisfactory condition of so many foreign countries, is delaying the movement ; but that it will come before long I am convinced, because the outturn of gold has increased very greatly during the past few years, and the richness of the South African gold fields promises that it will continue to increase for a considerable time to come. But the fall in gold which I look for will be the result of causes operating upon gold alone, and not of any bimetallic arrangement. Whether silver will rise is questionable. I am told that the processes of extracting silver ore from the substances with which it is found have quite lately been so much improved that the cost of production is materially reduced. I have no knowledge on the point, and I simply repeat what I hear. If it be true it is quite possible that silver may fall further ; at all events it is very improbable that there can be much rise. Therefore, even if I am right in the opinion that we are on the eve of a marked rise in prices, the divergence from the old ratio between gold and silver will remain nearly as great as ever. How do the Bimetallists propose to contract the divergence ?

I can think of only one way in which the value of gold can be greatly lowered, and that is by inducing the gold-using countries to greatly restrict their use of the metal and to take largely to silver. How do the Bimetallists propose to induce the gold-using countries to do this? The history of the present century proves that the world has gradually acquired a preference for gold and a dislike for silver, and that one after another the civilised nations have adopted the former as their standard and rejected the latter. Further, the experience of the United States from 1878 to last year shows conclusively that even where the Legislature, under the influence of sectional feeling, restores silver as a standard the capitalist classes, and more particularly the bankers, refuse to act upon the legislation and thereby defeat it. How, then, do the Bimetallists propose not only to persuade all the nations of the civilised world to agree to bimetallism and adopt the old ratio of $15\frac{1}{2}$ to 1, but also to prevent the capitalist classes, and more particularly the bankers, from rejecting the arrangement and insisting upon payment in gold? Have they ever asked themselves the question? Or are they even aware of the obstacles confronting them? The Bimetallists are never tired of telling us that bimetallism would have been adopted long ago, but for the opposition of England. Any single great State holding out, they admit, makes bimetallism impossible. Therefore, on their own showing, the United States must enter into the Bimetallist Convention if it is to be a reality. But the Constitution of the United States expressly declares the freedom of contract inviolable. Therefore, if the United States is to be a party to bimetallism, freedom of contract in the fullest sense must be maintained. Consequently, all the existing contracts for payment in gold would remain valid and enforceable, and everyone would be free in the future to enter into similar contracts. But to confine ourselves for the moment to our own country—most of the loans made in London are payable in gold ; not only State loans, but also company loans. If the United States, so many of the American railway companies, the central and southern American countries, most European Governments and India, not to mention China and Japan, have all to pay the principal and interest of their loans in gold, how is silver to be made international money? or how is the value of gold to be reduced by a bimetallic arrangement?

The truth of the matter is that the Bimetallists have never thought out the subject at all. They have taken up the extraordinary notion that the opening of the mints provides a market. When the error is pointed out they simply repeat that all debtors will pay in the cheapest money, forgetting that there are two parties to every contract, and that if it is an advantage to the debtor to pay in silver, it is a disadvantage to the creditor, and that, speaking generally, the creditor is the more powerful of the two. It is he who is asked to grant a favour, and it is he, therefore, who can impose the terms of the contract; while, as I have just been showing, the United States Constitution insures that every contract once made must be fulfilled in its integrity, or the defaulter can be declared a fraudulent bankrupt. I would advise the Bimetallists, then, to put aside their hardy assertions and airy predictions and to address themselves to a conscientious study of the difficulties of their own panacea. Firstly, the civilised countries have, with or without reason, acquired a preference for gold and a dislike for silver. How are those feelings to be overcome? Secondly, France and the nations of the Latin Union insist, if there is to be bimetallism, upon the old ratio of 15½ to 1 being restored. How is this condition to be complied with? Or what consideration can be offered to France and the other nations of the Latin Union to induce them to depart from the condition? Thirdly, the Constitution of the United States declares the freedom of contract sacred. Do the Bimetallists suppose that the people of the United States can be persuaded to change that plank in their Constitution? And if not, how can bimetallism be made real? Fourthly, the great majority of contracts in the civilised world have been concluded in gold, and, that being so, how can gold be made to cease from being the real money of the world and the real standard of value?

THE bimetallist delusion has its root in the belief that prices are regulated by the quantity of money in existence, meaning by "money" legal tender coin. I have in the foregoing papers endeavoured to show that this is absolutely wrong, and I need not go over the ground again here farther than to say that the Clearing House returns prove in the most convincing way that coin enters into payment for wholesale commodities only to a very trifling extent ; that the payments, except for a small fraction, are really made by cheques, bills, and other credit documents. It is to be observed that the economic discussion about prices has reference only to the wholesale prices of commodities. Retail prices are not in question ; partly because they differ not only from town to town, and from street to street, but from shop to shop. They depend largely upon custom, the laziness of housekeepers, their ignorance of business, their inability to judge the quality of what they buy. But the chief reason why retail prices are put aside is that they follow wholesale prices, haltingly and slowly, it is true, but still to some extent. If Bimetallists were clearer minded they would bear this fact in mind, and would recognise that in the wholesale markets coin plays so small a part that practically it has exceedingly little influence in determining prices. The more intelligent Bimetallists are compelled to admit that credit does affect prices as largely as coin. But even they try to belittle their admission. For example, Sir David Barbour asks whether the withdrawal of credit would have the same influence on prices as the withdrawal of gold. If Sir David attaches the same meaning to the word "withdrawal" in both cases, then I unhesitatingly answer, undoubtedly it would. Since the beginning of the present

---

* From *The Statist* of August 4, 1894.

year, to take an illustration, about 14 millions sterling in go'd have been withdrawn from the United States. The value of the exports from the United States immensely exceeds the value of the imports into the States; therefore there is an enormous balance of trade in favour of the United States. And yet gold is being withdrawn from the country at an alarming rate. Unless there is borrowing in some shape or other, by the Government by preference, it is clear that the gold drain cannot be stopped at present, much less that the gold withdrawn can be got back. Therefore the gold exported from the United States is for the time being lost to the country. If a similar amount of credit were as completely lost to the country, then it is as certain as anything can be that the influence upon prices would be precisely the same.

To show that this is so, I would refer Sir David Barbour to what happened in the City on the Baring collapse in 1890. The Messrs. Baring Bros. had carried on perhaps the greatest accepting business in the world, at all events one of the largest, and the crisis of November, 1890, caused the disappearance of the Baring acceptances. If Sir David Barbour will make inquiries he will find that the withdrawal of the credits which used to be afforded by the Messrs. Baring Bros. had quite as powerful an effect upon prices all over the world as the withdrawal of a similar amount of gold. Furthermore, all the great financial houses were compelled to reduce the credits they used to open to their customers and to accept upon a smaller scale, some of them because they had lost so heavily that it was not safe to continue on the old scale, others because the distrust was so deep and so general that they did not care to run risks on the old scale or felt that they themselves might suffer in reputation if they were seen to be doing business very largely. Therefore, the amount of credit given by London to the other countries of the world has been greatly reduced since 1890, and as a natural and inevitable result prices have come down lower than they have hitherto been in the present century. If Sir David Barbour and the Bimetallists generally could clear their minds from the delusion of the quantitative theory they would see from this and innumerable other instances that the withdrawal of credit has quite as much effect as the withdrawal of gold upon prices, and in fact that it is by restricting and compelling credit to be withdrawn that the withdrawal of gold has an influence upon prices at all. At times when the Bank of England's reserve is small and business is active, the withdrawal from the Bank of a million sterling would send a thrill of apprehension through the whole City, and would certainly have a paralysing effect upon the Stock Exchange, while in all probability wholesale prices would all come down. The lowest estimate I have seen of the gold holdings of the United Kingdom put them at about 70 millions sterling. A withdrawal of one million out of 70 millions would be, roughly, rather less than 1½ per cent. Can any sensible man really believe that if it were the quantity of gold which governs prices a decrease in the supply of about 1½ per cent. would have the effect upon prices which historically it is known such a withdrawal, and even a smaller withdrawal, has had upon markets in this country? The quantitative theory can thus be shown to be absurdly false by any number of arguments drawn from various sides. It can be disproved completely by a reference to the Clearing House returns, and it can equally be disproved by the experience gained from the withdrawals of gold from the Bank of England in past times. Similarly, what is going on in the United States at the present time shows the untenableness of the quantitative theory. The amount of gold that has been shipped from New York since the beginning of the year is large absolutely, no doubt, but it is not of commanding magnitude when compared with the whole volume of the currency of the United States. And yet not only in the United States but here in Europe the shipments are being watched with apprehension week by week, and people are asking anxiously, How will it all end?

One other evidence of the utter untenableness of the quantitative theory is afforded by the range of prices in the United States at the present time. Since 1879, when specie payments were resumed, the United States has retained a very large part of the gold raised at home, and it has imported immense amounts from Europe, and even from Australia direct. Even allowing for the withdrawals since 1890, the accumulation of gold in the United States has been very large. From 1878, when the Bland Act was passed, until 1890, when the Sherman Act took its place, the Treasury bought silver and coined it into legal tender dollars at the rate of two million dollars worth a month. From 1890 to 1893 the Treasury bought silver, paying for it in legal tender notes, at the rate of 4½ million ounces every month. In addition to these vast amounts of gold and silver, the United States has retained in circulation nearly 70 millions sterling of greenbacks, or Treasury notes issued during the war. And, lastly, there are about 24 millions of bank notes in circulation. There is thus an immense mass of currency in the United States—head for head nearly twice as much as in the United Kingdom, and even rather more than in France. And yet prices are not inflated in the United States. During the past twelve months America has exported immense quantities of all kinds of produce, and sold its exports at the current European prices—clear proof that the cost of production in the United States has not been so high as to prevent that country from competing with other countries in the European markets. It cannot be said that the money of the United States is depreciated. Prices, therefore, are on the gold basis because gold is the real standard of the country. And yet the currency is, undoubtedly, inflated because it has been artificially swelled by so much silver and notes. If the quantitative theory were true prices ought to be extravagantly high in the United States—so high that it would be impossible for the American producer to sell in the European markets. I do not suppose that those who have publicly committed themselves to the errors of bimetallism will be convinced by any number of arguments or illustrations or appeals to experience, let them be ever so convincing in themselves. But the great public who are desirous only to get at the truth will see, I hope, from what has been just said, that the quantitative theory, unsound in principle, is absolutely disproved by practice. An increase of the currency does not raise prices. A withdrawal of credit has precisely the same effect upon prices as a withdrawal of the standard metal; and, in fact, a withdrawal of the standard metal affects prices only because it affects credit—it is credit and not coin which regulates prices.

BIMETALLIST FALLACIES.*—V.

In contending that the Governments, if they all combined, could enforce bimetallism, the Bimetallists are practically maintaining that Governments can work miracles. Gold and silver are found in different parts of the world under very different conditions, and in very different quantities. They are mined at varying cost and they are esteemed in various ways. The demand for them varies greatly from time to time, the supply of them varies equally. In the time of Locke, silver was so universally the money of the world that that philosopher thought it the proper standard of value to be selected by Governments. Now silver has been rejected by almost the whole civilised world, and gold adopted as the standard of value. Yet, in effect, the Bimetallists tell those who listen to their lucubrations that Governments can overrule the laws of nature and of the human mind, and not only fix the ratio between gold and silver, but maintain it for ever and a day when once it is fixed. How reasonable men can so delude themselves by a jingle of words is one of the surprises of our time. "But," say the Bi-

* From *The Statist* of August 11, 1894.

metallists, "if all the Governments united to fix and maintain a ratio there would be no object in opposing. Debtors would gladly pay in silver, for bimetallism would lessen the burden of their debts, and creditors would readily accept silver." Is it, I would ask, because it would lessen the amount coming to them? "Oh, no," say the Bimetallists; "creditors are debtors as well as creditors, and what they would lose in their capacity as creditors they would gain in their capacity as debtors"; which is as much as to say that since all men in business have to buy as well as to sell, therefore, all men of business desire that prices should be low! Creditors, speaking generally, are not debtors to any serious extent; that is to say, their assets largely exceed their liabilities, and, therefore, it is to their interest that their return on their assets should be as large as possible. A gain upon the small liabilities at the expense of the large assets would be a very delusive gain indeed.

But is it true that if universal bimetallism were adopted people would have no motive for refusing to accept it? Clearly it is not true; the very opposite is true. Independent Governments cannot bind themselves for ever and a day. Supposing, for the sake of argument, that all the Governments could be induced to adopt bimetallism, it would be only for a term of years. All commercial Conventions of every kind are limited in time, and in the nature of the case must be so. Probably, the term selected would be ten years; that is about the time which statesmen generally think they are able to provide for. Every ten years, or so, the old generation of statesmen passes away, and a new one comes into existence, and with the new there come new conditions, new wants, and new feelings. But it does not matter very much what term of years might be decided upon. The Convention must clearly be terminable some time, and if it were terminable people generally would have the very strongest motives for refusing to accept silver. There would be the possibility, to say no more, that some of the nations might withdraw from the agreement. 'No," the Bimetallists say, " that would be impossible. The advantages of bimetallism would be so clear that if it were once adopted it would never be departed from." I would remind those who are so very cocksure on this point that Mr. Cobden was equally certain when he negotiated the commercial treaty with Napoleon III. that the advantages of low duties would be so plain to the humblest understanding that France would be converted to Free Trade, and that the conversion of France would be followed by that of all the world. I presume that even the most self-complacent Bimetallists will not assert that they are more prescient than was Mr. Cobden. We all know how completely the event has differed from Mr. Cobden's expectation; and it is surely possible, to say the least, that the expectation of the Bimetallists would be equally disappointed. But if the possibility of silver being again rejected was always before the mind of men of business, why should any reasonable man in the gold-using countries part with his gold and encumber himself with silver? The Governments in particular would certainly not do so. Preparing for war, they would conclude that gold would be always and everywhere acceptable, but that it was impossible to foresee what might happen with regard to silver. Therefore, they would go on in the future as they are now doing in the present, accumulating vast war treasures in gold. Similarly the bankers in all the gold-using countries would decide that they could not go wrong if they kept to their gold, but that they might be landed in the most disastrous consequences if they accepted silver to any considerable extent. Therefore, we may be very sure that all cautious bankers would refuse to run the risk, and that silver would be boycotted just at it was boycotted in the United States under the Bland and the Sherman Acts. Thirdly, I pointed out last week that according to the Bimetallists themselves bimetallism is impracticable unless all the great nations enter into it, and that if the United States enter into it the inviolability of contracts must be one of its cardinal conditions; therefore, all the gold contracts now existing would have to be observed. And what object under heaven would any reasonable man have in waiving his right to gold, and accepting a metal which, for at least the past twenty years, has been so grievously depreciated?

There is another and an exceedingly strong motive. I presume that even the most sanguine Bimetallists do not hope that differences respecting gold and silver would immediately come to an end if bimetallism were adopted. They know that gold is greatly preferred now all over the civilised world. They know that the vast majority of our own people are determined to keep to gold. and to have nothing to do with silver. They surely cannot delude themselves so completely as to hope that if the Governments by some extraordinary combination of accidents could be induced to adopt bimetallism all the preference for gold now existing would disappear. But if it did not; if large numbers of people continued to think that bimetallism was an absurdity, impracticable and doomed to inevitable failure, can Bimetallists persuade themselves that these people would not take care to insist upon all payments to them being made in gold? If the United States were a party to the arrangement the freedom of contract would have to be maintained; and, given freedom of contract, is it not certain that every banker, and every great capitalist, who has a preference for gold, and who disbelieves in bimetallism, would insist on payment being made to him in the metal he prefers? All who came to this country to borrow would, therefore, find that they were required to engage that the principal and interest of the loans granted them should be paid in gold. While the Sherman Act was in force one of the American Territories arranged with a syndicate in New York for a loan, and applied to Congress for power to conclude the arrangement. Congress insisted upon striking out the clause providing for payment in gold, and the consequence was that the loan was refused. Precisely the same thing would happen here, under bimetallism, supposing our Government went mad and adopted the system. Everyone who wished to safeguard the interests of investors would insist, when a foreign issue of any kind was to be made, upon a clause being put into the contract that payment of both principal and interest should be in gold. It would be easy to point out many other very strong motives for refusing to accept silver ; but I hope I have said enough to convince all people who are open to conviction that there is not the least chance in the world that bimetallis.n ever could be worked in this country, even if it were adopted everywhere else.

BIMETALLIST FALLACIES.*—VI.

SIR DAVID BARBOUR'S CONTENTION.†

SIR DAVID BARBOUR misses the point of the article to which he replied in last Saturday's *Statist*. My contention in that article was that if the word " withdrawal " is used in precisely the same sense in the two phrases " withdrawal of credit " and " withdrawal of gold," the withdrawal of credit has just as much influence upon prices as the withdrawal of gold ; or rather, to speak more correctly, the latter has an influence upon prices only because it reduces the volume of credit. Sir David also misunderstands my intention in referring to the shipments of gold from New York this year. Since New Year's Day over 14½ millions sterling in the metal have been exported from the United States, not because of an adverse balance of trade—for the value of the exports is immensely greater than the value of the imports—but because of the deep-seated distrust of the United States currency which prevails all over Europe. For the time being, therefore, the gold is absolutely lost to the American people. It may be got back, of course, by Government borrowing or by a sudden change in European opinion of the American situation. But for the

---

* From *The Statist* of August 18, 1894.
† See Appendix C.

time being the gold is lost. Now, if we are to compare the effect of the withdrawal of credit upon prices with the influence of such a withdrawal of gold as is going on from the United States, we must take an instance in which the credit also is absolutely lost for the time being. If the great London banks, for instance, were to become dissatisfied with the rates ruling in the Money Market just now, and were to curtail their loans and discounts by 14½ millions, everyone would understand that the credit was there all the same, that it was merely a struggle to get higher rates, and that people could get either loans or discounts if they complied with the terms of the banks. Such a withdrawal of credit is not comparable with the withdrawal of gold from the United States. But the destruction of credit by the Baring crisis *is* comparable, is exactly on all fours, with what is now going on in the United States. Then the credit absolutely ceased to exist. It was lost to this country and to all who were in the habit of being accorded it previously. And the result proves as clearly as anything can possibly do that a destruction of credit has as great an influence upon prices as a loss of gold.

Sir David Barbour shows, if he will pardon me for saying so, that he does not rightly understand the subject which he is treating of when he talks of the Baring crisis as the most marked symptom of a disease, but not the disease itself. Does he mean to imply, then, that the withdrawals of gold from the United States are the disease and not the most marked symptom of it? Surely he must know—indeed, his own language proves that he does know—that the Baring crisis was preceded by the most reckless speculation, and that the credit of the great house of Baring Brothers broke down thereby. Vague, general language about economic causes may cover a retreat, but it does not advance a genuine discussion. The simple fact is that Messrs. Baring Brothers had over-traded altogether, and abused their splendid credit so that a collapse ensued. It was a breakdown of credit in the clearest and most unmistakable way; and the crisis that has prevailed all over the world for the four years that have followed, shows the far-reaching consequences of such a breakdown of credit. Now let us turn to the withdrawals of gold from the United States. Are not they also symptoms of a disease? To speak in perfectly plain language, are they not the consequence of the breakdown of credit? The United States Government engaged in a wild silver experiment. It brought the currency of the country into utter disorder and, as in *The Statist* I ventured to warn my readers for over a year would surely be the case, a crisis followed. The gold shipments are a consequence of the distrust that ensued, or, in other words, of the breakdown of credit The real disease is the same in both cases—rash speculation of one kind or another, wild over-trading, financial experimentalising, general distrust, and paralysis of credit. It is the latter—the paralysis of credit—that immediately has an effect upon prices. Sir David Barbour professes himself unable to see what bearing the gold movements from the United States and the condition of the currency there have upon the argument in which we are engaged. I trust that those who are not so much committed to bimetallism will be able to understand it from this explanation of my position.

But, Sir David Barbour says, If it be true that the total circulation of the United Kingdom (gold and credit) amounts in round figures to 6,100 millions sterling, is it not clear that the withdrawal of the 100 millions of gold which are now held in the United Kingdom would have an infinitely greater influence upon prices than the withdrawal of 100 millions of credit? Except under the most extraordinary circumstances, it certainly is true. But it is true not because the quantitative theory of value is correct, but because the gold forms the banking reserves upon which the whole of the credits are based, and that if the banking reserves were to be dissipated the credits would rest upon no foundation. The 100 millions of gold which, according to the hypothesis, are held in this country are employed partly in the ordinary circulation and partly as banking reserves. That portion which is in the ordinary circulation has no more

influence upon wholesale prices, with which alone we are concerned here, than the token silver money or the Irish and Scotch
bank notes, which are not legal tender. It is the portion of the
gold which is employed as banking reserves that really affects
prices, and it affects prices only because it is a guarantee that the
credits rest upon a solid foundation. The gold in the banking
reserves in normal times is not employed. It is held as a reserve
for emergencies, and when an emergency arises all credit would be
utterly destroyed if no gold existed in the banks. Directly, then,
the gold in the banking reserves does not affect prices, because it
is held idle and is not intended to be employed in making payments. But its disappearance would cause consternation, because
every contract implies that if necessary the payment shall be discharged in the legal tender money of the country. It is very seldom that the legal tender money is required, but it must be
held by the banks as an assurance that it can be had if it is
really needed. To make his argument really relevant, Sir David
Barbour ought to have asked, not whether the withdrawal of
the whole of the gold in the United Kingdom would or would
not have a greater effect upon prices than the withdrawal of a
sixtieth part of the credit, but whether the withdrawal of all the
gold would or would not have a greater effect upon prices than
the destruction of the whole of the credit.

BIMETALLIST FALLACIES.*—VII.

IT is curious how difficult it is to make the ordinary man see that
the word " money " has a great many different meanings, and to
induce him to distinguish between those meanings in his reasoning. I have frequently pointed out in the course of these articles

---

* From *The Statist* of August 25, 1894.

that "money" is often used to signify loanable capital, as, for example, in such phrases as "the Money Market," "the value of money," and "the rates for money." In all these cases we do not speak of coin or cash or currency; what we really are thinking and talking about is loanable capital. The gold Monometallists, by confounding "money" in this sense with "money" in the sense of currency, are led into several of their mistakes; and the Bimetallists are equally in error. They frequently confound these two different senses in which the word money is employed. But even when we talk of money in the narrower sense—in the sense of coin, or currency, or cash—we ought to distinguish between the various uses to which it is put and the different functions it performs. And it is important to do so, because the influence of money upon prices is regulated by the function which the money performs. I have called attention once or twice already to the fact that money is sometimes hoarded up. Formerly, it was very common indeed for thrifty people to put away money, in actual coin, in an old stocking, or in a bed-tick, or in the seat of a chair, and so on; and in backward and semi-civilised countries the same thing is done still. Money when so hoarded is withdrawn altogether from the uses of trade, and can have no more influence upon the course of prices than the metal of which it is composed had while it was originally in the mines. Again, money is accumulated in war treasures, and while so accumulated it likewise is withdrawn from the uses of trade, and can have absolutely no influence upon the course of prices. Thirdly, money is used in the ordinary circulation of the country; and fourthly, it is used as bank reserves. Few people who think at all can be blind to the fact that money when hoarded, or locked up in a war treasure, is out of the service of trade, and therefore can have no influence upon prices. But the Bimetallists generally labour under the delusion that money circulating amongst the public affects wholesale prices, whereas it does not.

The gold sovereigns and half-sovereigns which circulate in this country are doing precisely the same kind of work as bank notes

and silver and copper token coins. They are, in fact, doing the
work of token coins, and have absolutely no influence upon prices.
I have referred again and again to the fact that in the wholesale
markets—with which alone we are concerned in economic dis-
cussions about prices—money in the form of coin and bank notes
enters only to a trifling extent; that payments or settlements are
really made by means of bills and cheques and other credit instruments. The sovereigns and half-sovereigns which circulate
amongst the public are used to a large extent, no doubt, in paying
wages; and to a still larger extent in the retail trade. Most men
who have business of any considerable importance make their
payments by means of cheques. It is chiefly women who carry
gold sovereigns about, unless, as already said, where they are required for paying wages; and in these cases they are used for the
convenience of the employers, not for that of the employed. In-
quire into the fact, and you will find that almost as soon as he
receives it, the ordinary labourer changes the sovereign paid him
as wages—for it is seldom that a single purchase of his amounts
to so much. It stands to reason that the man who earns little
more than a pound a week must make his purchases in small
amounts, and consequently must require to change the sovereign
almost at once. Therefore, the sovereigns and half-sovereigns
used for paying wages are used for the convenience of the employers, partly to reduce the bulk of the coin which has to be
carried from the bank, and partly to save labour in counting out
the money. The sovereigns so used, then, simply take the place
of shillings and half-crowns and florins, and they do so because it
is more convenient for the employers. The amount of money
that is thus actually in circulation is not in the least influenced by
the course of prices; it is determined entirely by the numbers of
people employed, and by the rate of wages. When trade is active
and employment very full, wages are usually high, and therefore
there has to be an increase in the circulation, simply because the
amount of money paid weekly by employers to their workpeople
is larger than usual. That active trade and good employment

lead to a rise in prices is true enough ; but they do this not because the circulation is larger, but because the working classes earn more than at other times, and consequently have more to spend. The Bimetallists, however, are unable to see that money in circulation —what we may call currency—has no effect upon prices ; that it is used almost entirely in the payment of wages and in the retail trades, and consequently that it does not require to be legal tender. A one-pound note in Scotland or Ireland is just as effective in the payment of wages and in purchases in the retail trades as the sovereign ; and yet the one-pound note is not legal tender. If, then, a mere increase of the currency, as the Bimetallists fondly dream, would affect prices, it is folly of them to clamour for so costly a material as silver. Paper would serve their purpose equally well, and would be far cheaper for the State.

The only money which really does affect prices is that which is contained in the banking reserves. That must have a legal tender character, for the banks must pay in legal tender because their reserves are called out only when their own credit is more or less in danger. A bank deals in credit, and it is through its credit that it operates upon prices. But it has to safeguard that credit by keeping a reserve upon which it can fall back in times of emergency. In the first place, then, the money constituting the bank reserves must be legal tender, whereas the money in the currency of the country need not be so. It is mere waste to circulate such a mass of sovereigns and half-sovereigns as are circulated in this country when one-pound notes would do as well without the legal tender character. Further, the money constituting the banking reserves must have a value of its own, absolutely independent of the stamp of the Government or of the credit of the institution issuing it, if it is to be a real money and is to maintain its value in all seasons of crisis. Where specie payments are suspended, notes issued either by the Government or by the State bank take the place of real money to a certain extent. But almost invariably these notes get depreciated, and though they have to be used by the banks for their reserves they are not real money all the same.

Wherever the banking system is sound, and the country has *bona-fide* money, that part of it which constitutes the banking reserves must not only be legal tender, it must have intrinsic value ; whereas the money circulating amongst the public need have neither a legal tender character nor intrinsic value. A paper note issued by a bank without privileges of any kind would serve just as well as gold money, provided the bank had really good credit—as, for example, the Scotch banks have. In short, the only real money a country need have is that required for the banking reserves. And a country with an ideally perfect banking system and with a due sense of the wisdom of economising cash would have no other standard money. If money was so economised ; if all the banks of the civilised world kept adequate reserves and money was not wasted in the circulation, then there would be plenty of gold for the world's purposes, and prices would be regulated by the ebb and flow of credit.

THE ECONOMIC INFLUENCE OF THE WAR PREPARATIONS.*—I.

LORD FARRER'S letters to us lately and the bimetallist agitation combine to show that the public generally has failed to grasp the economic influence exercised by the war preparations of the great military nations. Certain aspects of this influence have been noted again and again, but the full measure seems to have escaped general observation. And yet it is clear that these preparations have exercised a paramount influence on the world's development for a quarter of a century. It may be worth while, therefore, to call special attention to the subject, and to point out in a little detail

---

* From *The Statist* of September 8, 1894.

some of the principal ways in which the influence has made itself felt. Firstly, ever since France set herself resolutely to reorganise her army after the disasters of the struggle with Germany there has been an intermittent fear of a great European war. In 1875 the danger of a renewal of the conflict between Germany and France became very real—indeed, was only averted by the intervention at Berlin of Russia and the United Kingdom. During the Russo-Turkish war the danger was equally imminent. The Penjdeh collision again brought war within measurable distance. And the Schnäbele incident once more excited alarm all over the Continent. For some few years now apprehension has been less keen, and the confidence of the business community in the maintenance of peace has been undoubtedly greater. But for all that there is a latent dread that any accident may precipitate a general conflict. While this feeling has lasted for over twenty years now it is inevitable that enterprise should be held in check. For a very long time the fear was so great that few people cared to engage in any business that could not be concluded in less than a year, and the consequence was that many important undertakings which would have conduced to the well-being of the world had to be postponed. Business lived from hand to mouth, which meant, of course, that profits were small and activity slight. Furthermore, year by year more and more of the male population of the Continent is being with-drawn from productive enterprise and kept under arms during the most impressionable period of life. The production of the world thereby has been curtailed. If the immense masses of men kept under arms since the Franco-German war had been employed in industry, they would have produced something or other that would have exchanged for the things that have in fact been produced. Thereby they would not only have added directly to the wealth of the world, but they would have made more profitable the enter-prises that have been carried on. And, thirdly, the necessity for maintaining such vast masses of men in unproductive employ-ment has wasted the world's resources.

Fourthly, the millions of men kept under arms the vast expenditure upon munitions of war of all kinds and on fortifications, and the immense stores that have been accumulated and renewed have compelled the Governments to increase enormously the taxes and to pile debt upon debt, thereby further augmenting taxation. Italy furnishes the most striking illustration of the consequences. Of course, the financial difficulties of Italy are not due to military preparations alone. Very many other causes combine—corruption, bad administration, extravagant outlay upon public works, banking mismanagement and dishonesty, reckless speculation and unwise protection have all contributed powerfully. But allowing for all that in the fullest degree, the excessive armaments maintained have undoubtedly weighed very heavily upon the kingdom. Russia and Austria-Hungary are suffering less, but they are suffering much for all that. Germany is better able to bear the strain than the other countries mentioned, but the burden on Germany is still very great. Even in France, with all her wealth and all the thrift of her people, the military policy is felt grievously. The smaller States have been compelled to act more modestly; but there is no doubt at all that the strain upon the smaller States, too, is very great. The Spanish army and navy are out of proportion, for instance, to the resources of the country, and the Balkan States—Roumania, Servia and Bulgaria—have to keep up far greater armies than their development warrants, simply because in no other way can they hope to maintain their national independence. Fifthly, the waste of inventive genius during the past quarter of a century has been lamentable. If the ingenuity which has been devoted to the invention of quick-firing rifles, of smokeless powder, of artillery of great range and piercing power, of armour to resist the heaviest metal, of bullet-proof garments, and the like had been employed in introducing some great new industry, or in perfecting labour-saving machinery, how much would the well-being of the world have been promoted !

Perhaps, however, the way in which the military preparations have told adversely upon the economic development of the world most profoundly is seen in the accumulation of vast war treasures.

For some time before the Franco-German war the production of gold had been falling off. The decrease was still more marked in the Seventies, and in the first six years of the Eighties it became more decided still. While this was going on several of the leading nations of the world demonetised silver and adopted the gold standard. Under any circumstances, therefore, gold would have become scarce and the prices of commodities would have fallen. But just while the ·production of gold was declining in the most marked way, and while the nations were competing with one another for the metal for purely monetary purposes, the great military nations began to withdraw it in vast amounts from the service of trade and to lock it up in Treasuries and State banks as a precaution against the dreaded European struggle. It is singular how little most of those who have been writing and talking about currency during the past twenty years have given attention to this most remarkable phenomenon of our time. Yet nobody who studies the matter closely can doubt that the accumulation of war treasures has intensified to a most serious degree the fall in prices, and so has deepened the malaise, the unrest, and the discontent which are everywhere surging around us. There are many contributory causes, no doubt, of Socialism and Anarchism. But personally I have little doubt that the war preparations are among the most potent, and that of these again the war treasures are not the least operative.

For example, Russia has accumulated about 100 millions sterling of gold. The Austro-Hungarian Bank holds nearly 13 millions sterling. These two sums are clearly not bank reserves. The latter—the Austro-Hungarian—may become a bank reserve by-and-by, but at present it is not so. When we come to France and Germany it is impossible to say how much the war treasures amount to. The German Government holds six millions sterling in the fortress of Spandau; but how much of the gold in the Imperial Bank of Germany, and in the Bank of France is a war treasure no one can say; but that a large part is everyone knows. Passing to the smaller States, it is notorious that all of those look

upon part of the gold held as a war treasure. It is certain, therefore, that the war treasures in Europe alone amount to something
between 150 and 200 millions sterling.

THE ECONOMIC INFLUENCE OF WAR
PREPARATIONS.*†—II.

THERE is published in *The Statist* to-day a letter signed "X," in
which the above estimate of the war treasures in Europe is called
in question. If I were at liberty to give the name of our correspondent, he would be recognised as one of the most eminent men in
the City, who speaks with special authority on the subject of which
he writes. He undoubtedly has opportunities for learning the
intentions of the Treasuries and State banks all over Europe, and
therefore what he says is entitled to great weight. But in the
absence of his exceptional information I see no grounds for
modifying my estimate. I speak, of course, with due deference,
and admit that in this matter "X" is better placed to know the
truth than I am.

As regards Russia, it has always hitherto appeared to me incredible that the very able men who of late years have administered
the finances of that country are under the delusion that in locking up gold which nobody can get at, they are helping to raise the
value of the rouble. If they are, it is another illustration of a
phenomenon, common enough yet always surprising; namely,
the union of very weak reasoning faculties with great administra-
tive capacity. A promissory note is equal to gold when it can be
exchanged for that metal at the option of the holder. But the
holder's estimate of the worth of the note is not in the least
affected when he is told that the issuer has plenty of gold to
redeem it, but intends to take good care not to pay away the

---

* From *The Statist* of Sept. 22, 1894.
† See Appendix D.

metal. The Russian Government has a very rich gold field. How can it believe that gold in the Imperial Bank, in the Treasury, and with its agents abroad raises its credit more than the gold in its own mines, or the vast resources of which it disposes ? The real reasons why the credit of Russia stands so high are the steady growth of the population, the immense resources of the Empire, the rapidity with which it has recovered from the famine of 1891, the scrupulous care with which it has always fulfilled its pecuniary obligations, the readiness of France to lend what it requires, and so on. At the present time the Russian Government has gold enough to resume specie payments if it so chose. But it does not choose, and therefore I submit with deference that it shows that the gold is not held at the service of trade. From the purely military point of view the Russian Government is right in its action The Czar is well known to be intent upon peace; but he is making every possible preparation for war if it is to come, and amongst his preparations not the least important is the accumulation of such a vast amount of gold. If war were to break out, and the gold was not held, the Russian Government would have to borrow at the very outbreak of hostilities. But it is extremely probable that it could not borrow either in Germany or in London ; and it would hardly be politic to borrow in France, for France, on the hypothesis, would have need for large amounts of gold, and it would not be expedient, therefore, to call upon her to supply Russia as well as herself. The 100 millions sterling now held are ample for all probable requirements. Every expenditure at home could be paid for in paper ; it is only abroad that gold would be needed, and before so large a sum as 100 millions sterling was spent abroad, in all likelihood the result of the war would be determined. Either Russia would have gained such advantages that she would be able to levy requisitions and in large measure to make the war self-supporting, or she would have been driven back into her own interior provinces. In the latter case, the possession of a large amount of gold would simply tempt the victor to impose a ruinously heavy indemnity. I submit, then, that the policy of the Russian Government is clear, that it has been well thought out, and that it is admirably adapted for the purpose in view.

Turning next to Austria-Hungary, I venture to think that its motive in accumulating so large a reserve is equally evident. The army of Austria-Hungary is by no means equal to that of either Germany, France, or Russia. Her fortifications are incomplete, her population is not homogeneous. In the event of a great war her credit would hardly be so good as that of the three other Powers named. Yet if she did not possess a large war treasure she would have to borrow at extreme disadvantage. But the Government of Austria-Hungary is pre-eminently cautious. It might suit a Bismarck to tell the world frankly that he was accumulating gold as a precaution against war, but neither the Emperor Francis Joseph nor his advisers are likely to be as candidly cynical. To them, a pretext would strongly commend itself, and what better can there be than the resumption of specie payments? But I must say, with deference, that it seems to me incredible that such prudent and able men would choose a time of feverish war preparations for resuming specie payments. If the war cloud passes away no doubt resumption will take place; if it bursts, the most critical cannot blame the Austro-Hungarian Government for suspending the operation and returning to inconvertible paper. Coming, in the third place, to France, I admit the force of "X's" argument. The Bank of France undoubtedly requires a very large reserve, not only because of the magnitude of its note circulation and the extent of French trade and French investments abroad, but still more because of the depreciation of silver and the largeness of the silver circulation in the country. It is, therefore, exceedingly difficult to estimate how much of the gold reserve of the Bank of France is held for those purposes, and how much as a war treasure But, so far as the world knows, the French Government holds no special war chest; therefore, as in 1870, on the breaking out of a great war, France would have to borrow, and in the first place she would be compelled to draw largely on the Bank of France.

" X " may have reasons for what he says unknown to me, but I submit that the management of the Bank of France is far less admirable than I believe it to be if it has not taken the eventuality into account, and accumulated a large war treasure. Nobody knows better than " X " the obstacles put in the way of considerable withdrawals of gold from the Bank of France. Are not those obstacles a convincing proof that the Bank desires to hold something more than a mere adequate banking reserve? Passing, in the last place, to Germany, I would observe that the six millions sterling held in the fortress of Spandau as a war chest would go but a very short way in case of such a great war as the world fears. Germany would need very much larger sums. And is it in accordance with the proverbial foresight of the German Government in military matters to leave itself dependent upon any foreign country for the funds it would need? Is it not as certain as anything in the future can be that Germany likewise would have to borrow on the outbreak of hostilities, and that she would turn in the first place to the Imperial Bank? The Imperial Bank of Germany is not so completely a branch of the Finance Ministry as is the Imperial Bank of Russia. But it is to be recollected that the Imperial Chancellor is in the last resort the arbiter of the policy of the Bank.

While, then, I am glad to have the opinion of " X " upon this extremely important matter, and while I admit the authority with which he speaks, and the special knowledge he is in a position to obtain, I am unable to see where I have over-estimated the European war treasures. At all events, I recommend readers to peruse carefully " X's" letter, and at the same time I present to them the grounds on which I based my own estimate.

THE INDEBTEDNESS AND CURRENCY OF
THE UNITED STATES.*

In a recent article in *The Statist* it was shown that in the first
nine months of the current financial year of the United States—
the nine months that ended with March—the value of the
exports of commodities from the United States exceeded the value
of the imports of commodities by somewhat more than 44½
millions sterling, that the value of the exports of silver exceeded
the value of the imports of that metal by somewhat over 5½
millions sterling, and that, adding the silver and the commodities
together, the value of the exports of both during the nine months
exceeded the value of the imports also of both by, roughly, 50¼
millions sterling. And yet gold is being exported in considerable
amounts. It may be objected that about 10 millions sterling in
gold were borrowed in the autumn during the height of the currency crisis, that the loans have now fallen due and have to be repaid. Let us, then, put the figures in a somewhat different way.
The exports of silver and commodities together exceeded in value
the imports of silver and commodities by 50¼ millions sterling.
But 10 millions sterling of gold were borrowed and imported, and
have to be repaid. The excess of the exports of treasure and
commodities over the imports of the same was therefore, in round
figures, 40¼ millions sterling. This would be at the rate of very
nearly 54 millions sterling for the whole year. Is it credible that
the indebtedness of the United States to Europe amounts to 54
millions sterling per annum? And if it is not, how does it happen that an excess of exports over imports at the rate of nearly
54 millions sterling in a single year does not enable the United
States to settle without exporting gold the special debts incurred
during the crisis last autumn of somewhat over 10 millions sterling? The question at first sight is not easy to answer; and yet it

---

* From *The Statist* of May 19, 1894.

is difficult to believe that the indebtedness of the United States to Europe amounts to 54 millions sterling per annum—or, to express the matter more correctly, that the interests, dividends, profits, and commissions annually payable by the United States to Europe amount to about 54 millions sterling per annum. Even at 5 per cent., this would represent a capital debt, or a capital investment, or whatever phrase may be preferred, of nearly 1,100 millions sterling. It may be said that Europe has been investing in the United States ever since the colonisation of America began, and that it is not at all improbable that the capital investments now represent about 1,100 millions sterling.

But let us look at the matter a little more closely. At the present time, as readers have frequently been reminded in *The Statist*, about one-third of the total railway mileage of the United States is in the hands of Receivers, and consequently an immense amount of both interest and dividends that used to be paid is now in default. Further, a large number of the mortgage companies are insolvent. Silver mining has been contracted in the most serious manner, farming is in a depressed state, trade is paralysed, credit is in a deplorable condition. Clearly, then, it follows that the interest, dividends, and profits that can be paid this year must be enormously less than they were in the good times that preceded 1890. Furthermore, it is notorious that ever since the middle of 1890 European capitalists have been withdrawing capital on a very large scale from the United States. Therefore the investments of Europe in the United States must be greatly less now than they were four or five years ago. If allowance be made for the bankruptcies that have occurred, for the depression that is universal, and for the withdrawals of capital during the past four years, it seems not to be an exaggerated estimate that, assuming 54 millions sterling to be payable in the present year from the United States to Europe—representing a capital value of 1,100 millions sterling—the capital value of the European investments four or five years ago cannot have been much less than 2,000 millions sterling. But 2,000 millions

H

sterling is greatly above the highest estimate that has been framed
of the total capital of Italy. It is about a fourth of M. de
Foville's estimate of the capital of France, and it is about a
fifth of Mr. Giffen's estimate of the capital of the United Kingdom. Is it credible that the European investments in the United
States exceed the total worth of Italy and everything it contains,
and amount to about one-fifth of the total capital of everything in
the United Kingdom? I confess that I find it very difficult to believe such a thing. The explanation of the figures I would offer
is that the withdrawal of capital which has been going on ever since
the difficulties of the Barings became serious is continuing upon a
very large scale, and that the excess of exports over imports represents, therefore, to a very considerable extent, capital withdrawn.
I need not remind the reader that there is extreme depression
all over the United States, that it is very difficult, therefore, to
employ money to advantage, and that consequently there would
not be anything surprising if capitalists were to withdraw capital
in the hope of finding better outlets for it in Europe or elsewhere.
Further, it is to be borne in mind that there is extreme distrust.
President Cleveland had influence enough to get the Sherman Act
repealed; but the Inflationists have not lost courage. They tried to
pass the Seigniorage Bill over his veto, they have been endeavouring
to carry a Free Coinage Bill, and it is known that they will seize
the first opportunity to push forward some other measure of a
similar character. Besides that, the Democratic Party seems to
be hopelessly demoralised. One fraction of it is trying to discredit the President, and to defeat his policy, and the delay in
carrying the Tariff Bill is causing widespread dissatisfaction. For
all these reasons—and I might add many others—it is natural
to suppose that capitalists are withdrawing capital. But there is
another reason, and it is the redundancy of the currency.

In the issue of *The Statist* of September 23rd last year it was
shown that on the last day of August, 1893, the total paper circulation of the United States outside of the Treasury—the paper,
that is, in banks and in the hands of the public—amounted to

about 203 millions sterling, composed of greenbacks, Treasury notes issued under the Sherman Act, silver certificates, and bank notes. There was, in addition, a considerable amount of gold coin and about 12 millions sterling of silver coin. It is quite evident that this enormous paper circulation is far in excess of the requirements of the country. If trade were very active and speculation rampant, it is possible that employment might be found for the greater part of it. But it is perfectly clear that in the present condition of the United States the circulation is altogether redundant. And it is a well-established fact that where the currency is redundant the worst kind drives out the better. The greenbacks, the new Treasury notes, the silver certificates, and the bank notes, are not receivable as money abroad, but they are receivable at home. Gold coin is, however, receivable everywhere. Naturally, therefore, the paper is driving the gold out of circulation ; and, as a matter of course, it is the foreign capitalists who are most sensitive to this state of things, and are withdrawing their capital as a precaution. The New York correspondent of *The Statist*, in the telegram published from him a fortnight ago, states distinctly that the gold withdrawals were not affecting markets ; and he is a well-informed and careful observer. I accept his word without hesitation to this extent: that the general public and the ordinary operators are paying no attention to the gold exports, looking upon them as made in the natural course of trade. But I venture to think that it is altogether different with the foreign capitalists; that they are by no means sure that the United States will be able to maintain the gold standard, and that, therefore, they are taking timely precautions. Nobody need be surprised at the uneasiness of foreign capitalists in view of the enormous excess of the exports over the imports pointed out above. It has has been shown above that, even allowing for the 10 millions sterling of gold borrowed last autumn, the value of the exports of all kinds exceeds the value of the imports of all kinds by over 40 millions sterling, being at the rate of nearly 54 millions sterling for the year. Now, if it can be held that this represents the annual, normal, regular in-

H 2

debtedness of the United States to Europe, it is not surprising
that the more thoughtful capitalists should come to the conclusion
that it is impossible for the United States to maintain the gold
standard. In the piping times before 1890, Europe invested upon
such an immense scale in the United States that the indebtedness
of the latter was masked. Now Europe is not investing—is, on
the contrary, withdrawing capital—and the foreign capitalists
argue that the real facts are therefore being disclosed. I cannot believe that this is the true interpretation. During the
five years ended with 1890 the exports of gold exceeded the
imports by less than 17½ million dollars, or about 3½ millions
sterling. Is it credible that if during those five years the indebtedness of the United States to Europe amounted to 60 or 70
millions sterling per annum the shipments of gold could have
been so small? in other words, that the indebtedness of the
United States to Europe could have been almost covered by the
excess of goods exports over imports, and by European investments in the United States?

THE OUTLOOK IN THE UNITED STATES.*—I.

IF it be true that foreign capitalists are withdrawing the money
they have invested in the United States, and that domestic
capitalists, as asserted by American papers, are sending money
away to Europe for safe keeping, then the position is growing
very serious indeed. The banks are afraid to supply exporters
with the gold they require. The exporters are drawing upon
the Treasury, and as a consequence the reserve in the Treasury
is being rapidly depleted. It will be in the recollection of
the reader that on February 1st the Government invited ten-
ders for 10 millions sterling, bearing 5 per cent. interest, the
issue price being 117. The loan was taken in full, and yielded

* From *The Statist* of June 9, 1894.

the Treasury nearly 12 millions sterling. Barely four months have elapsed and the Treasury is once more in a position in which it urgently requires to borrow. It would be absurd to go on in this way. The Treasury is getting gold from the banks and out of the circulation, and is accumulating it in a form so handy for export that it is accordingly being exported as fast as it is collected. The result must be, unless there is a complete change in the public opinion of the country, an enormous increase of the debt charge and the disappearance of gold from the circulation. It seems useless to expect the proper remedies from the present Congress. President Cleveland could be depended upon to take the right measures if Congress would follow his lead. But apparently he has lost influence with Congress, and therefore is helpless. Still, it may be worth while to point out the true remedy. It is manifest that those who are expecting permanent improvement from the passing of the Tariff Bill will be disappointed. Some kind of settlement of the tariff question is desirable. It will do a little towards improving matters, but it will not do much. So again, a temporary loan would for the moment produce a better impression. But neither a Tariff Act nor a temporary loan will avail much; the true remedy is quite different.

It is perfectly clear that the circulation of the United States is redundant, and that the redundant paper currency is driving out gold. The effectual remedy, then, to apply is, in the first place, to reduce the paper currency. The time no doubt is very unfavourable. But as the opportune moment has been allowed to pass by there is no use in dwelling upon that. What is necessary has to be done, whether under favourable or unfavourable circumstances. What ought to be done immediately is to withdraw the greenbacks from circulation and cancel them. They amount to nearly 70 millions sterling, and if they were all withdrawn they would reduce the paper currency so much that it is to be hoped confidence would be restored. This is on the assumption that the American Government and the American people have made up their minds to maintain the gold standard at whatever cost.

If they have, then the paper currency ought to be reduced. The February loan of 10 millions sterling at 5 per cent. calls for an annual interest of half a million sterling. Now it seems as if another loan of 10 millions sterling would be required. If that loan is raised, then the interest on the debt will be increased by a million sterling annually and nothing really effectual will have been done. Gold will have been taken from the banks and from the circulation, and accumulated in the Treasury in a position in which it can be most easily sent out of the country. The evil, in fact, will after a while be intensified instead of allayed. If the Government were to borrow 70 millions sterling at 3 per cent.— I presume there would be no serious difficulty in so great a country as the United States borrowing 70 millions sterling at 3 per cent., or very little more—that would involve an annual interest of less then 2¼ millions sterling, or, adding on Sinking Fund, the whole additional charge for the debt would still be only between two and three times as great as the Government will, probably, have to incur even if it only borrows from hand to mouth as occasion requires. The advantage of a large loan would be that it would constitute an effectual remedy, while temporary small loans do no permanent good. The greenbacks were issued during the war. They are Treasury notes, pure and simple. They were not issued either against gold or against silver. But the Government is bound to redeem them in gold; and it has pledged itself to keep a reserve of 100 million dollars, or 20 millions sterling, always in the Treasury to insure their convertibility. These greenbacks are legal tender for all amounts; they are held largely by the banks in their reserves, and, in fact, the bank notes are redeemable in greenbacks. Their withdrawal would not merely reduce the volume of the paper currency very greatly; it would make a void in the bank reserves, which would have to be filled up, most likely with gold.

If the Government were to adopt the measure which I have now been recommending, no doubt it would have a serious effect for a while upon the European Money Markets. A loan of 70 millions sterling would have to be issued in Europe as well as in America, and gold would be taken. That, of course, would deplete the stocks in Europe, and would upset the Money Markets for a time. But the withdrawal of the greenbacks would be a slow and gradual process; therefore, the paying up of the instalments of the loan could be spread over several years. The disturbance of the Money Markets, if that course were adopted, would therefore be minimised, and as the production of gold is increasing very rapidly, there is no reason to fear that the United States Government by borrowing the amount required would permanently injure business. There are large stocks of gold in many countries where they are not now required—in India, for example. And probably much of the gold that has been accumulated by Austria-Hungary would be got back again if the United States set itself in proper manner to restore order in its finances. But, however that may be, the question for the United States Government to consider is not so much the influence of its policy upon other countries as the effect of that policy upon the prosperity of its own country. The American Government and the American people of course can, if they please, let their gold go and adopt the single silver standard. But if they will not do that, and are resolved to maintain the gold standard, then it seems obvious that the redundant paper currency must be reduced. The withdrawal of the greenbacks, however, though it would be a very important step in the right direction, and would have a powerful influence on public opinion at home and abroad, would not be enough. It should be followed by a reform of the banking system. That, however, is a subject too large to be dealt with at the tail-end of an article like the present.

THE OUTLOOK IN THE UNITED STATES.*—II.

In the foregoing articles attention has been called to the evidence showing that capital is being withdrawn from the United States by Europeans and Americans alike, because of a fear that the country will have to come, after all, to a silver standard. And I have pointed out what, I venture to think, is the true remedy. It is to be feared, however, that there is very little likelihood of the true remedy being adopted ; indeed, it would almost look, incredible as it may appear, as if the majority in Congress rather wishes for another crisis. Last year public opinion in the United States declared so emphatically and so unequivocally against the Sherman Act that even the Senate, stubborn as it was and independent as it felt itself, had at last to give way, and the Sherman Act was repealed. But the majority in Congress, though it bowed to the will of the people, under the pressure brought to bear upon it, did not let the matter end there. On the contrary, the Silver Seigniorage Bill was passed very shortly afterwards, though it was vetoed by the President; and an attempt was made to carry a Free Coinage Bill. The attempts in these two cases proved to the majority in Congress that the President would neither be intimidated nor cajoled ; that he would not allow of any tampering with the currency. Therefore the majority resolved not to give the President the means which he required for tiding the country over its difficulties. From that day to this the majority in Congress has refused even to consider a Bill authorising the President to borrow a sufficient amount at a low rate of interest. From all this it looks as if the majority resented the pressure brought to bear upon it by the public last year, and had resolved to punish both the public and the President by refusing to give the latter

---

* From *The Statist* of June 23, 1894.

the means of carrying out his wise policy, and by allowing matters so to drift as to bring on another currency scare. Many persons argue, indeed, that the majority is calculating that the public would then turn round and say "The Sherman Act has been repealed, yet the currency difficulty is as great as ever; therefore let us try whether a Free Coinage Bill will do."

It seems incredible that this can be the settled policy of the representatives of the American people, for it is as certain as anything can be that if the policy is carried through and another scare is caused, the crisis will be far worse than that of last year. Last year the authority of the President was unimpaired. He was backed by public opinion, and there was no serious doubt that after a while the Sherman Act would be repealed. But now the influence of the President with his own party seems to be altogether undermined, and nobody can foretell what the majority may do. Therefore, if there is another scare, gold will first go to a premium, will then be hoarded, and practically will disappear from circulation. If that happens, the Treasury will be unable to meet its obligations in gold. If gold disappears, great as is the credit of the United States Government, and vast as are the resources of the country, it will be impossible for the Treasury to find what does not exist. Consequently, it will have to cash the greenbacks in silver; and if matters come to that there will be such a fall in the prices of securities as has not been witnessed for many a long day. Remember that silver at present is actually worth less than half what it was worth a quarter of a century ago. Therefore, if the greenbacks, which amount to very nearly 70 millions sterling, were to be paid in silver at the present value of that metal they would be worth less than 35 millions sterling. Similarly with the new Treasury notes. And it is obvious that the fall upon the Stock Exchange would be almost unprecedented; for if silver became the money of the country all railway rates would be received in that metal or its equivalent, and so, also, would all railway dividends be paid in that metal. Such a panic, therefore, as would be caused has rarely been witnessed in the history of the world; and it is incredible that a legislative assembly elected by an intelligent people like the American, can be purposely manœuvring to bring about such a catastrophe. For myself, I do not believe that the majority has the intention that is sometimes attributed to it; but that there are people who suspect it of the intention is beyond dispute.

Assuming, however, for the sake of argument, that Congress does refuse to take any of the measures requisite to restore confidence; that, therefore, distrust deepens, until at last gold goes to a premium, is hoarded, and ultimately disappears from the circulation, it is clear that the panic which would ensue would be unparalleled in its magnitude. And yet it is not to be doubted that the prosperity of the United States would be very little affected; that after a while business would adapt itself to the new conditions, and then the country would go on prospering just as before. After all, the greatness and wealth of a country depend upon the qualities of its people and the extent of its resources, not upon the kind of money it employs. If gold were to disappear from the American circulation it would flow over to Europe, and the supplies in Europe would be enormously increased just at the time when the output from the mines is augmenting so rapidly. There would consequently be a marked rise in prices in Europe, and silver, like everything else, would share in the advance. Furthermore, the disappearance of gold from the American circulation would cause the currency to be very scarce. It is redundant now; but the export of the gold would end the redundancy and cause scarcity instead. Consequently there would be a demand for silver for coinage, supposing that silver became the real money of the country. Moreover, silver being worth only about half what it was twenty years ago, if prices were measured in silver they would have to be about doubled, and therefore the quantity of silver required for the circulation in America would be very large; firstly, because the gold displaced would have to be replaced, and secondly, because prices reckoned in silver would be so much enhanced that the amount of silver required would be very considerable. I am inclined to think, therefore, that the demand for silver for the United States, and the addition made to the gold supplies of Europe, would raise materially the value of silver all over the world. It is highly probable, moreover, that the example of the United States and the rise in silver would very soon compel the Indian Government to reopen its mints. There would, in consequence, be an inrush of silver into India that would be certain to further send up the price. Lastly, the example of so great, so enterprising, and so prosperous a country as the United States would be likely to be followed by all backward countries in the world. Every country with disordered finances, that could at all manage to maintain or to resume specie payments, would be likely to be greatly impressed by the action of the United States, and therefore the use of silver would in all probability be immensely extended. It is by no means clear therefore, whether silver would not be gradually raised near to its old value if it were to be adopted as a standard by the United States, especially as the additions to the European supply of gold, both from the United States and from the mines, would be such as to send down considerably the value of gold. If gold were to tend downwards and silver to advance rapidly upwards the meeting point might not be so far off as is now supposed.

GOLD TO THE UNITED STATES?*

THERE is a vague notion that there will be a drain of gold to New York in the autumn. But a drain of gold to New York in the autumn is highly improbable. Of course, the Government may borrow in gold as it borrowed last February. Of course, also, the great railway companies may borrow gold in London, as they did last autumn, and either action may lead to a considerable drain of gold to New York. But if there is not borrowing of gold in

* From *The Statist* of August 25, 1894.

some form or other, a drain of the metal to New York seems very unlikely in the autumn. The recent rains have delayed the harvest, but the crops for all that are looking exceedingly well, and we have reason to hope that the harvest, not only at home, but generally speaking all over Europe, will be good. Therefore, here is nothing to suggest an extraordinary European demand for wheat from the United States by-and-by. Even if the harvest should turn out bad we may safely conclude that Russia, India, Australia, Argentina, and all the various other countries that supply us will compete keenly with the United States, and therefore that America will not have such a control of our wheat market as will enable it to take an extraordinary amount of gold. Again, it does not seem probable that the British investing public will rush wildly to buy American railroad securities. The British public may go mad; but it is not probable, and if there is not a great speculation in American railroad securities there will be no drain of gold because of British investments in the United States. For the condition of that country—the paralysis of trade, the want of employment, the misery and unrest of the working population— all combine to discourage investment in any form. But if it is not likely that we shall buy immense quantities of American grain, or American securities, or that we shall invest largely either in industries, or in lands, or otherwise, how is there to be a great drain of gold? Lastly, it is to be presumed there is not likely to be such an outburst of activity in our own manufactures that there will be an immense import of raw materials. Without borrowing, then, I confess myself unable to see how the drain of gold is to arise.

While the disorder in the American currency lasts, it is to be feared that distrust of all things American will continue in Europe, and that, therefore, shipments of gold from America to Europe are far more likely than shipments of the metal from Europe to America. Over and over again we have been assured by American journalists of all kinds that the gold shipments were practically at an end, and week after week the assurances have been contradicted by events. Now our own wiseacres are predicting that what is being lost at present by the United States will be taken back again in the autumn ; and the prediction is just as well founded as were the assurances of the American Press just referred to. As has been pointed out in *The Statist*, over and over again during the past two years and more, the silver experiment of the United States has thrown the currency into hopeless confusion. The currency is redundant, and, as always happens when that is the case, the worse form of currency is driving out the better. Gold is receivable everywhere. Gold, therèfore, is being sent from the United States, and is being accumulated in Europe in immense quantities. Much of it came first to London, now it is being taken by Germany and Austria ; and although the export may cease for a while, every now and then, it is reasonably certain that it will begin again. Of course, I freely admit that even yet it is possible to apply a remedy. Indeed, there seems no room for doubt that if Congress was amenable to reason, and would follow the leadership of President Cleveland, order would be by-and-by restored. But nothing seems less likely than that Congress will listen to reason, or follow the President. In the House of Representatives the majority is disciplined enough. The House was elected directly by manhood suffrage at the same time as the President. The House, therefore, is in full harmony with him, and knows that those who voted for it wished it to support, in the fullest way, the President chosen together with it. But only one-third of the Senate was elected at the same time. Another third was chosen four years ago, and the remaining third has now been sitting for very nearly six years. Two-thirds of the Senate, there-fore, was chosen at a time when the feeling of the country was altogether different from what it is now, and it is not surprising in consequence that it should be out of harmony with the President and with the House. The result, unfortunately for the United States, is that there is a dead-lock between the two Chambers. In the Senate, indeed, a portion of the Democrats are almost as hostile to the President as the Republicans themselves. But if

Congress will do nothing I greatly fear that matters must become worse before they can be better.

The danger, then, is real that the country will have to come to a silver standard. I do not say that it *must* come—that would be to predict—but only that there is danger that it will come. As has been observed more than once already, the real prosperity of a country depends not upon the money it uses, but on the qualities of its people and its material resources. Therefore, in the long run, it does not seriously matter what standard of value the United States adopts. For all that, passing from the gold to the silver standard would mean a crisis such as is rarely seen, and temporary chaos. It is possible that when the danger reveals itself fully to the American people they will insist upon Congress taking the necessary action. Last year the Senate was just as much opposed to the repeal of the Sherman Act as it is now to a reasonable tariff reform. And yet popular opinion compelled it to assent to repeal. So it is at least possible that if the public becomes thoroughly alarmed it will insist upon the proper remedies being adopted by Congress. Even if the public is confused and unable to trace the root of its difficulties, and, therefore, is as undecided as Congress itself, it is just possible that the banks may combine to supply the country with a currency that will serve the occasion. As has been often noted in *The Statist*, American bankers are well trained in the practice of combining for what they believe to be either useful or necessary. It was the boycott of silver by the banks which defeated the silver experiment forced on by Congress ; and every reader will recollect how, last year, the banks combined to support one another, issuing Clearing House certificates, and accepting them as if they were legal tender. Therefore, when the occasion arises it is possible that the banks may combine to supply the country with a currency that will be generally accepted. And there are other ways, no doubt, in which the crisis may be turned or averted. But, unless in some way or other a remedy is applied, the danger is very great indeed that the country will have to come to a silver standard. That being so, European capitalists will have lost their cunning if they send their gold in the autumn in large amounts to the United States. And similarly, the American capitalists must act very differently from their kind if they take exceptional measures to force gold into a country in which there is only too much reason to fear that gold, before long, will rise to a premium, and will disappear altogether from the circulation. What seems most likely is that continued shipments of gold from New York, interrupted possibly from time to time, but always renewed, will go on until either the currency is restored to order, or gold altogether disappears from the circulation.

THE DIFFICULTIES OF THE PRODUCING COUNTRIES.*

THE difficulties of the producing countries, which are made so much of by the Bimetallists, have nothing in them mysterious or even puzzling ; the causes are plain enough. In good times the more backward countries largely work by means of capital raised in Europe, and more especially in this country. Investors are then anxious to employ their money profitably, and they are ready to engage in all kinds of new enterprises. Those who wish to push new enterprises forward or to extend old enterprises therefore flock from all parts of the world to London. Funds are raised by means of the public issue of loans and companies, by attracting deposits, by borrowing on security from banks, and in short in every conceivable way. As long as this goes on the backward countries grow rapidly in prosperity, the area under cultivation is extended, railways are pushed in new directions, public works of all kinds are constructed. But there comes a check when credit receives a shock in Europe. During the years immediately preceding the Baring crisis the amount of capital invested by this country abroad was unprecedentedly large. But

* From *The Statist* of July 7, 1894.

as soon as the Baring embarrassments became serious, that capital began to be called home. Some of the great financial houses were compelled to bring home their capital in the hope of sustaining their position ; others were alarmed at the possible consequences of the crisis they saw to be impending, and they also withdrew money. As the crisis deepened, the apprehension spread, and more and more investors called home their money. When the crisis actually occurred it will be in the recollection of our readers how great was the alarm here, and how long it continued. Nobody knew what other houses might come down ; credit for the time was paralysed; and the first thought of all who were engaged in business was to strengthen their own position. The withdrawal of capital then became more eager than ever. Soon new causes of uneasiness sprang up. The bankruptcy of Argentina, the wild speculation that followed the revolution in Brazil, the civil war in Chili, the default of Portugal, the increasing financial embarrassments of Spain, Greece, and Italy intensified the scare, and more and more capital was withdrawn. The refusal of this country to lend more to the Australian Colonies brought on a crisis there ; and as it was seen to be approaching depositors withdrew their funds in immense amounts from the banks. Meantime it became clear that the Sherman Act was a failure, and that it would involve the United States in serious embarrassments. Silver fell ruinously, and alarm sprang up respecting the silver-using countries all over the world. This caused the withdrawals of capital to continue on a still greater scale.

This long-continued withdrawal of capital threw the more backward countries into the most serious difficulties. They would have been embarrassed by the mere stoppage of further supplies. As already said, they have been developed in the past largely by means of capital supplied from Europe, and the refusal of Europe to place more funds at their disposal would have prevented them from continuing their development at the old rate. But when Europe began not merely to refuse fresh supplies, but to insist upon the repayment of large portions of what had already been advanced, their embarrassments became grave indeed. In many cases Governments had paid the interest on their debts only by means of fresh loans. Everyone can now see that this was the case with the South American countries, with Portugal, Greece, Spain, and to a lesser extent with Italy, with our own colonies of Australia likewise. The inability of the Governments to get further assistance compelled them to draw upon the banks. First they withdrew the deposits they had with the banks ; then they proceeded to borrow from the banks. In doing this they reduced the funds at the disposal of the banks ; and when the banks found themselves in danger of being unable to repay the deposits that were being withdrawn they called upon their debtors to pay up. The consequence was severe and widespread distress and numerous failures. The debtors, to save their credit, were compelled to sell whatever there was a market for. The first step in the process was a great fall in Stock Exchange securities ; the next was a similar fall in commodities. The man who sees bankruptcy staring him in the face cannot wait for better prices ; he has no option but to take what he can get for his goods ; and so the pressure on all the debtors throughout the backward countries to sell brought about the low prices over which so much moan is made. The difficulties of the debtors abroad were of course intensified by the shock to credit all over Europe, and particularly in this country. Business of all kinds shrank here. Very many were unable to buy ; very many more were unwilling to do so. Most people were far more anxious to protect their position than to venture into new undertakings. Thus the producing countries were under the necessity to sell as best they could, while the investing countries were afraid to buy upon any great scale. A fall of prices was then inevitable. When the Baring collapse, occurred, I pointed out in this journal that there must be a great shrinkage in our foreign trade and that this must lead to lower prices. Any reader who will look at *The Statist* of November

I

22, 1890, will see how fully my forecast then has been verified by the event. What it was possible to foresee clearly four years ago is not, then, in its nature either mysterious or puzzling; on the contrary, what we are now experiencing is a natural consequence of the over-trading of the four or five years before 1890, and the crash that then followed.

I am far from denying that the appreciation of gold has intensified the fall in prices. I admit fully that there is an appreciation of gold, and in this I differ from the gold Monometallists, who, in my opinion, are injuring their own cause by refusing to acknowledge what to me appears to be an incontestible truth. But while in my opinion the gold Monometallists are wrong upon this point, and the Bimetallists are right, I am equally convinced that the Bimetallists so grossly exaggerate the influence of the appreciation of gold, and so misapprehend its bearing upon the economic condition of the world, that they put themselves altogether out of court. The appreciation of gold, of course, aggravates the difficulties of the backward countries which have borrowed too much gold. But it only aggravates; it does not create the difficulties. Whether gold had or had not appreciated, the countries which borrowed too much gold would be embarrassed. If they borrowed greatly more than they could afford to pay, they would have to default sooner or later. But the appreciation of gold by adding to the burden of their debts necessarily increased their embarrassments. If Argentina, Portugal, Greece, Spain, Italy, and India had been wisely governed, had kept their borrowings well within their means of payment, their Governments could look with indifference upon the appreciation of gold. The real cause of their embarrassments is that they were not wisely governed, but rushed into the most reckless extravagance. It may be admitted further that the difficulties of the producing countries have been aggravated in another way by the appreciation of gold. Prices are lower than they otherwise would be, because of that depreciation. But, as has been pointed out above, prices must necessarily have fallen ruinously whether there was or was not appreciation of gold, because the withdrawals of capital by Europe from the producing countries have been on so enormous a scale during the past four years that those countries have been compelled to sell what they could dispose of at whatever prices they could get. And, unfortunately, the disposition in Europe was to buy little and to grant no credit. The crisis, then, through which the world has been passing for the past four years is the immediate cause of the embarrassments of the producing countries. As the crisis passes away, their condition will begin to improve ; and such of them as have recuperative energy will become prosperous once more when Europe, and more particularly when this country, is ready to invest in those countries on the old scale.

ENGLAND'S FINANCIAL PREDOMINANCE *

STRIKING evidence is being afforded just now of the ascendency of this country in the economic system of the world. Since the Baring collapse distrust has prevailed here more or less continuously. The public has lost confidence in its old financial leaders, and therefore is unwilling to invest abroad as it used to do formerly. Consequently, as explained above in the article upon the difficulties of the producing countries, European capitalists, and more particularly British capitalists, have been withdrawing capital on an unparalleled scale from all the newer countries. To this is due the banking crisis in Australia last year, and the general breakdown of the Australasian Colonies. To a considerable extent also the breakdown in South America is traceable to the same cause. And, lastly, the difficulties of the United States have been immensely increased by the feeling of distrust that has existed here. Moreover, the collapse of so many continental countries is a

* From *The Statist* of July 14, 1894.

direct result of the unwillingness of the British public to invest. From all this it is clear that it is essential to the modern economic system of the world that this country should lend largely. When it does, the economic machine works smoothly; when it does not, there is every kind of disorder. It almost looks, in fact, as if only two countries in the world at present are in a position to save much more than they require for their own immediate needs, those two countries being England and France. Almost all other countries—though some of them save on a very large scale—yet want more capital than they are able to lay by themselves. Holland and Belgium are small communities, and though they are rich and thrifty, yet the amounts with which they can accommodate other countries are not of much account. Germany has made wonderful progress during the past thirty years, indeed during the past half century; but yet it looks as if Germany requires almost all her own capital, and when she invests abroad she has to lean more or less upon this country.

France confines her investments to a few countries. Her condition differs very greatly from our own. Everyone in France saves to some small extent; but the number of great fortunes is few. The savings of France, therefore, consist of a very large number of very small sums; and the small people, being timid and rather restricted in their knowledge of the world, do not care to invest widely. When they *do* invest, however, their power is remarkably great, as has been shown recently by the success with which they have rehabilitated the credit of Russia. Furthermore, the greater financial power of France than of Germany is clearly proved by the difficulties of Italy. If France had not sold Italian securities on a great scale the embarrassments of Italy would be much smaller than they are. Even now, if France were willing to come to the help of Italy, it is almost certain that her worst embarrassments would be surmounted. Germany was quite willing, for political reasons, to aid Italy. But the purse of Germany is not as heavy as that of France, and French hostility, therefore, has proved more powerful than German help. We ourselves have not cared to meddle much with Italian Finance, and therefore Italy is in her present predicament. Practically, it may be said, then, that French financial aid is confined to Russia, Turkey, Greece, Spain, Portugal, and Egypt. Within its own limits it is very powerful; but the limits are narrow, and the timidity of the French investor prevents France from exercising a great influence upon the world generally.

Our country differs in almost all respects from France. Great fortunes are much more numerous, and our banks are much more powerful as well as far greater in number. Our banks, too, are scattered all over the country. They gather up the savings of every class, they transfer those savings to London, and thus accumulate them at the centre in a form in which they can be employed with the greatest possible effect for any given purpose. Then our great financial houses are much more enterprising, much more closely connected with all parts of the world than are those of France; and the result is that because of all these peculiarities, and because of the world-wide character of our trade, we have become the bankers for the whole earth. Whenever our investors are inclined to go into new enterprises, promoters flock to London from all quarters under the sun. Companies and loans are brought out with bewildering rapidity, and vast sums are raised as if there was no trouble in saving money. As long as this willingness to invest lasts prosperity spreads from country to country, and in the process gathers momentum. But as soon as our investors decline new enterprises there is a halt, and when our people begin to withdraw capital the newer countries are thrown into difficulties. It is the unwillingness of our investors to engage in new enterprises since the Baring crisis that has really brought about all the difficulties which have been experienced since. And if the Bimetallists would turn away from their craze and study the real facts of the situation, they would see that it is the want of confidence here, not the want of money, that is weighing upon the world of trade. The truth appears to be that England is a much richer country than is generally supposed, that she employs much more capital abroad than anyone has had an idea of hitherto, and that the great works that are being constructed all over North and South America, South Africa, our Colonies and India are really and truly being carried on by means of British capital, British skill, and British enterprise. The prosperity of the newer countries, then, is in one sense of the word deceptive. It is not a domestic prosperity to anything like the extent that is generally believed; it is rather the result of our own boldness and our own enterprise.

THE LOW PRICES.*

MANY observers are puzzled because, in spite of the great increase in the production of gold, prices have continued to fall so that they are actually lower now than they were when the gold output was at the minimum. The circumstance would indeed be puzzling if the quantitative theory of prices were true. But it is only another illustration of the falseness of that theory. It has been pointed out again and again in the foregoing papers, and dwelt upon with emphasis, that prices are really regulated by credit, and that credit is only to a small extent affected by the amount of money, or of gold, in existence. The total volume of credit that can exist at one time is limited, of course, by the magnitude of the bank reserves. But while the maximum of credit is so limited, the actual volume of credit at any moment is determined by a multitude of causes quite independent of the largeness or smallness of the bank reserves. Just now the volume of credit is limited, not by the quantity of the bank reserves, but by the after-consequences of the Baring collapse. The Baring collapse immediately compelled the great financial houses to restrict to an unusual extent their acceptances. Some did this because they were themselves more or less discredited; some did

* From *The Statist* of October 27, 1894.

it merely because they did not wish remarks made on the amount of acceptances of theirs outstanding. When the great financial houses restricted the accommodation they gave to their foreign customers in this way, their foreign customers were unable to do the same amount of business as before, and one result was that the newer and the poorer countries were plunged in difficulties. The whole of South America has suffered more or less ever since. So has the United States. So likewise has Australia. I need hardly add that the distrust in London, and the consequent unwillingness to lend, has largely contributed to the embarrassments of Greece, Portugal, Italy, and Spain. The distrust in London having thus plunged so many countries in difficulties, those difficulties in their turn reacted upon London. The great financial houses here at first were unwilling to lend because of the distrust existing at home; and to guard against what might follow the distrust, they withdrew large amounts of money employed by them abroad.

The refusal of new credits, and the withdrawal of old credits, embarrassed the countries largely financed from London; and as the embarrassments of those countries grew, London became timid of giving credit. Thus the whole volume of credit throughout the world has been enormously lessened since the Baring crisis, and the lessening is one of its results. Until the volume of credit expands it is impossible that prices should rise. On the contrary, every contraction of credit has necessitated a fall in prices. The newer countries, which produce food and raw materials for the older countries, have been compelled to sell at whatever prices they could obtain, because they could not get credit in London, and because they were compelled to meet their obligations. But as distrust at home prevented speculation, there was a paucity of buyers, and buyers not being very eager, held aloof in the hope of getting better and ever better terms. So the embarrassments of the producing countries, and the distrust of the capitalist countries, have made it more difficult to carry on transactions, and at last sales have been effected only by continually reducing prices.

There is nothing mysterious or puzzling, or incomprehensible in all this when it is looked at from the right standpoint. And it is perfectly clear that no increase in the gold production could have any material influence upon what has happened. Gold is employed only to a very small fractional extent, either in international dealings or in the wholesale markets for commodities within the same country, and the mere additions made to the amount of gold could consequently have no effect upon prices. If credit had been good the increase in the gold output would have enabled the banks to lend and discount more freely than before ; and every addition to the loans and discounts would have been an addition to the volume of credit existing, and so would have tended to send up prices. But as credit was bad, the increase in the gold output has simply gone to swell the bank reserves and the war treasures ; but has had no influence upon prices. Credit will revive under other influences than the results of gold-mining. Indeed, credit has already somewhat recovered, and will recover still more by-and-by. But at present the recovery goes no farther than this : that capitalists of all kinds do not feel it necessary to hold immense sums unemployed as they did until the other day, but are investing in sound securities, especially in home securities. There is an inclination, too, to invest largely in South Africa, and there is a desire to find other fields of investment, which promise to be both safe and profitable. In short, there is more readiness to encourage new enterprise than there has been for the past four years. But there is no doubt that the revival of credit is being checked by the political apprehensions that exist, and also by the still serious embarrassments of so many foreign countries.

APPENDIX A.

SCARCITY OF GOLD.*

SIR,—You scarcely do justice to the argument which gold Mono-metallists found on the state of the gold reserves and the rate of discount.

It is quite possible that appreciation of, or growing scarcity of, gold may accompany low rates of discount; it may even, as you point out, be in some cases the ultimate cause of low rates of discount. It is also true that an increase in the quantity of gold may accompany, or be the cause of, high rates of discount.

As a matter of fact, the rate of discount depends upon the demand for money, as compared with the supply of money ; and in this relation the demand is the predominant and most varying factor, whilst the quantity of gold available for the time being is, generally speaking, of quite minor importance.

You may not agree with me in this statement; and we will, therefore, confine our attention to those special cases in which, as we both agree, an increase or diminution of the quantity of gold available may become an important factor in the Money Market, and in its effect on prices. Your hypothesis, in which I agree with you, is, that the nexus, and the only nexus, between an increase or diminution in the quantity of gold for the time being available for currency, on the one hand, and wholesale prices on the other, is to be found in the operation of that increase or diminution on the state of the bank reserves, and on the consequent lowering or raising of the rate of discount. If this hypothesis is sound, it follows necessarily that an increasing scarcity of gold cannot have an effect in lowering prices without first operating to diminish gold reserves, and to increase the rate of discount.

---

* From *The Statist* of June 30, 1894.

If, therefore, during the period of low prices which has continued during the last twenty years or more, we find that the supply of gold in the bank reserves has been as abundant as ever, and that the rates of discount have not been higher, and have not fluctuated more than they did in previous years, and if it cannot be shown that lowering of prices has been preceded or accompanied by a diminution in the gold reserves, or a raising of the rate of discount, a strong presumption, to say the least, is raised that the low prices have not been due to an increased scarcity of gold.—I am, Sir, &c.,

June 27, 1894.   FARRER.

APPENDIX B.

GOLD AS A MEDIUM OF EXCHANGE, AND PRICES.*

SIR,—You are right in supposing that I tried to be short and became obscure. It is, in truth, impossible to do justice to this abstruse and difficult question in such a compass as either you or I can afford. But there are two points to which I should like to advert.

Your argument is, that the quantity of gold available for reserves has not increased in proportion to the business of the world. This argument seems to me to be founded on a fallacious assumption ; viz., on the assumption that there is some fixed proportion between the amount of business and the quantity of gold needed for the conduct of that business, and that this proportion remains constant through different times and in different countries. I believe that this assumption is contrary to fact. Changes in the modes of doing business, and the growth of credit, make it possible, and even necessary, for an ever-increasing amount of business to be done safely and properly upon the basis of an

---

* From *The Statist* of July 14, 1894.

ever-diminishing proportion of gold; and the most advanced countries do their business, and do it in the best way, with an ever-diminishing proportion of gold as a basis.

With the advance of commerce the demand for the precious metals as media of exchange is proportionately an ever-diminishing demand.

It would be trespassing on your space and on your readers' patience to give facts and figures to prove so elementary a proposition as this. Any history of trade which describes the growth of credit, any tables of statistics which give the amount of trade carried on, and of precious metals used as currency, at different times and in different countries, would establish and illustrate it.

But, even putting aside considerations such as these, there was, as a matter of fact, no depletion of the gold reserves of the world during the period which the supposed depletion of those reserves is said to have lowered prices. On the contrary, they increased considerably.

Dr. Soetbeer, in his " Materialien," 1886, p. 70, has given us an estimate or account of these reserves for several years (1877–85), and they are as follows :—

	Million marks.			Million marks.
1877	2,890	1882	4,070
1878	2,580	1883	4,600
1879	3,500	1884	4,880
1880	3,790	1885	5,040
1881	3,900		

There was, therefore, during the period in question, no reduction in the bank reserves of gold ; but, on the contrary, a large increase.

Dr. Neuman Spallart, in his " Uebersichten," 1887, p. 454, after pointing out that in the previous fifteen years the whole circulation of the civilised world, including gold and notes, had not much altered in amount, says :—

" The special characteristic of this last period is that enormous quantities of the precious metals have been collected in the treasuries and reserves of the banks ; the increase of dead and unproductive treasure from 1876 to 1885 may be reckoned at more than 2,400 million marks (£120,000,000); and from these facts, coupled with the increased use of commerical and banking credit, it is clear that the supply of money (Goldstand) in the markets of the world was an abundant one, and the rates of discount sank to what would before have been considered an impossible minimum."

So far as concerns the actual supply of gold for currency purposes. But there is another part of the argument of gold Mono-metallists which you do not touch. Your own hypothesis is that the way, and the only way, in which a diminution in the quantity of gold available for currency can affect prices is through the rate of discount. When gold becomes scarce, the rate of discount rises, enterprise is checked, and prices fall. This, at any rate, is your own hypothesis, in which I agree with you so far as to think that when the scarcity of gold as a medium of exchange does affect prices, it can affect them only by raising the rate of discount.

The question then arises : " Is there any evidence that the alleged scarcity of gold has raised the rate of discount ? " To this the answer is emphatically " No." The rate of discount has not risen more or been more fluctuating since 1872, which is the date fixed for the commencement of the supposed gold scarcity, than it had been before ; on the contrary, it has been lower and has fluctuated less.

In the Appendix to the Third Report of the Depression of Trade Commission, pp. 370 to 374, and in Dr. Soetbeer's " Materialien," Part VI., we have a statement of the rate of interest or discount for a series of years. From these tables it appears that the average rate of discount of the Bank of England was :—

For the five years ending	1865	...	...	...	...£4	17	6
,, ,, ,,	1870	...	...	...	... 3	11	7
,, ,, ,,	1875	...	...	...	... 3	14	10
,, ,, ,,	1880	...	...	...	... 2	18	3
,, ,, ,,	1885	...	.	...	... 3	8	2

The records of the Banks of France and Germany tell a

similar story of increased reserves, with no higher rates of discount.

If, again, we look to the fluctuations in the rate of discount—which, even more than the actual amount of the rate, tells us what have been the needs for gold and the apprehensions of the loss of gold (the hoisting of the danger signal, as it has been called)—we find, indeed, that in 1873 there was a quite unusual number of fluctuations—viz., twenty-four—in the course of the year. But this is just what we should expect, considering that in 1873 came the collapse after the unprecedented inflation of the previous years, the first great German demand for gold, and the payment of the French indemnity.

If, however, instead of taking the single year 1873 we take the thirteen years 1873 to 1885 inclusive, and if we compare them with the thirteen years preceding 1873—i.e., 1860 to 1873 inclusive—we find that the average number of fluctuations in the rate of discount in the Bank of England was 10 per cent. per annum in the years 1860 to 1872, and only 8½ per cent. in the years 1873 to 1885.

More striking than any figures is the diagram published by Bates and Hendy, Walbrook, of the rates of discount from 1836 to 1885, giving the changes month by month in a form which strikes the eye. Anyone who looks at this will, I think, be satisfied at a glance that, if the rate of discount affords any evidence of an abundant or a deficient supply of gold, there is nothing in the rates for the period subsequent to 1873 to show that there has been any less abundance of gold, or that there has been any greater difficulty in retaining or recalling gold during that period of alleged comparative scarcity and of low prices, than there had been in the previous period of alleged comparative abundance and of high prices.

If, then, the alleged scarcity of gold as a medium of exchange has not operated to raise the rates of discount it cannot have operated to lower prices. This is the Monometallist argument, and it appears to me to be a very strong one, and to be untouched by what you say.

But I despair of being able to do justice to it or to the causes which affect the rate of discount in such a letter as this, and beg to refer you to the enclosed pamphlet, " Gold Credit and Prices," pp. 57 to 79 (Cassell, 1889), in which I have stated the case with all the clearness and terseness of which I am capable. You will find that I have given full credit to the various factors which raise or lower the rate of discount, amongst which I believe the quantity of gold available as a medium of exchange to be comparatively unimportant.

Nor do I plead guilty to any confusion of mind on the subject of that ambiguous word " money." " Money," in the ordinary City sense of the word, is really " loanable capital," of which cash or gold forms a very inconsiderable proportion. If, indeed, credit were to fail altogether, our money or medium of exchange would be limited to the precious metals, and " money " would in that case become synonymous with " cash " or " gold." But this is an imaginary case, and the wildest panic is only a distant approach towards it. I am disposed to retort the charge of confusion of mind on those who confound " money " with " cash," and who, by adopting in one form or another the exploded quantitative theory of gold and prices, have done more to advance bi-metallism than any of its professed advocates.

The above reasoning is confined entirely to the effect of a supposed scarcity of gold considered as a medium of exchange. Considered as a standard of value, no reasonable man doubts that gold follows the usual laws of value, and that if the world's aggregate demand for gold for all purposes has grown in a greater proportion than the world's aggregate supply, the value of gold, including the £ sterling, must have risen, and gold prices must *pro tanto* have fallen. But has this been proved ?—I am, Sir, &c.,

FARRER.

Abinger Hall, July 11, 1894.

APPENDIX C.

THE QUANTITY THEORY.*

Sir,—In an article in *The Statist* of 4th inst., signed by Mr. T. Lloyd, I am represented as asking whether the withdrawal of credit would have the same influence on prices as the withdrawal of gold, and Mr. Lloyd answers the question unhesitatingly in the affirmative.

Mr. Lloyd probably refers to some remarks of mine in *The National Review* for July last, in which I argued against the theory of those who hold that the quantity of money has practically no effect on prices, because credit has, quantity for quantity, the same effect on prices as money, and because the amount of credit enormously exceeds the quantity of money.

The exact language which I used is as follows :

" The first objection is that the quantity of money has practically no effect on prices, or, as it was put by one authority, that there are £6,000 millions of credit in England, and only £100 millions of gold ; and that any possible addition of gold to the £6,100 millions of circulating medium (credit and gold) in England would have no perceptible effect on prices. Volumes might be written on the relation between prices and the quantity of money, but the argument to which I have alluded may be disposed of very briefly. I do not know how the figure giving the quantity of credit in England has been obtained ; but if the circulating medium of England is composed of £6,000 millions of credit and £100 millions of gold, and if credit and gold have quantity for quantity, precisely the same effect on prices, the reduction of credit by £100 millions would have no perceptible

---

* From *The Statist* of August 11, 1894.

effect on prices. Could the same thing be said if £100 millions of gold were withdrawn, or £80 millions, or £60 millions?

"I think that any such withdrawal would produce not merely a greater effect than the withdrawal of an equal amount of credit, but that it would shake our financial system to its foundation. . . . .

"Again, let us assume that £6,100 millions of gold are added to our circulation. According to the theory about credit and gold, which I have just stated, this increase in the total circulating medium should double prices ; yet the man who holds that such an increase of gold would merely double prices is not a person with whom it is worth while to argue.

"The cases which I have stated are, of course, extreme cases ; but it is extreme cases which afford the readiest test of the soundness of a general law."

The arguments which I have just re-stated appear to me perfectly sound when used for the purpose for which I employed them ; namely, to show the utter absurdity of holding that, quantity for quantity, credit and gold have precisely the same effect on prices. My opinion as to the relations between money, credit, and prices is not shaken by anything Mr. Lloyd has said.

Mr. Lloyd appears to think that the fact that gold is being exported from the United States, although her exports exceed her imports, proves that credit has as much effect on prices as gold. I do not know what bearing this assertion has on the point at issue. The export or import of gold is generally, but not invariably, determined by the balance of indebtedness for the time being, and not merely by the relation between exports and imports. In any case, I do not see that the export of gold from America proves anything as regards the relation between money, credit, and prices.

The Baring collapse of 1890 was preceded by many foolish investments, and much rash speculation in the financial world. It was the most marked symptom of a disease, but it was not the disease itself. I see nothing to be gained by contrasting the alleged effects of a financial and commercial crisis, arising from the operation of economic causes, with the probable effects of a mere contraction of the circulating medium, even though the crisis bring with it, among other results, a loss of confidence and contraction of credit.

Neither can I see that the quantity theory is disproved, because the United States has silver certificates, greenbacks, and bank notes in circulation, and nevertheless the gold standard is effectively maintained in that country. Mr. Lloyd states that the currency of the United States is "undoubtedly inflated, because it has been artificially swelled by so much silver and notes."

The circulation of a country may have additions made to it artificially by silver and notes to a very great extent, without destroying the metallic basis of the standard.

What is quite certain is that if the inflation is carried far enough the gold standard will cease to exist. This is the very result which a writer in *The Statist* of August 4th appears to anticipate in the United States, unless special measures are adopted. See the article headed "The Coming Danger in the United States."

To my mind the phenomena which we now observe in the United States furnish an argument in favour of the quantity theory.—I am, Sir, &c., DAVID BARBOUR.

Dalrannoch, Comrie, N.B., August 8th.

APPENDIX D.

WAR TREASURES.*

SIR,—In your leading article of last Saturday you call particular attention to the withdrawal from the service of trade of the very large amounts of gold which have accumulated during the last twenty years in the Treasuries and State banks of the leading con-

---

* From *The Statist* of Sept. 22, 1894.

tinental nations, and you express surprise that so serious a matter has attracted very little notice.

I share your surprise, and I fully agree with you that this withdrawal, however caused—during a period which, in spite of enormous outlay upon national armaments, has been one of very great financial and industrial expansion—has had a most injurious effect, not only upon prices of commodities, but upon general prosperity. As I write, the reports of a strong demand in the open market for gold for Germany suggest that the process of accumulation on the Continent has not yet come to an end.

But when you ascribe this locking up of gold so largely to the formation of war chests we part company. Your view does not seem to find support in the evidence before us. That military considerations have entered into the calculation I cannot doubt, nor that the leading European nations have established reserves of gold to meet sudden outbreaks of war ; but that any such amount has been set aside for war chests as 150 to 200 millions I have seen no reason for believing. The bulk of the accumulations in question can be otherwise accounted for.

In the case of Germany, Austro-Hungary, and Italy changes in their currency have necessitated the acquisition of most of the gold they have obtained. Russia is on a paper basis ; but for a number of years her Finance Ministers have been striving, by accumulating gold, to improve the national credit and to raise the value of the rouble, in which they have succeeded. France has closed her mint to silver without altering her standard of value, but it appears to me she has had ample reason for wishing to hold more and more gold in the fact that her circulation of silver is very large and that the gold value of the metal in this coinage has been declining until it is now only one half of its nominal amount.

In addition to the causes I have referred to, is it not natural that nations which see silver demonetised and discredited should endeavour to secure and to retain increased supplies of gold, the appreciating metal ? When you have made due allowance for these influences upon the continental demand, the amount of gold in the war chests will, I think, be found very far below the figures at which you have put it.

Whether you accept my explanation or not, there remains the serious matter on which we are already agreed, that a very large amount of gold has been practically withdrawn in Europe from the ordinary service of the world, or has, at any rate, been rendered difficult of access, which has operated to general disadvantage. Is this state of things to continue ? If so, how long will it take to fill up the void thus created ? Production of gold is indeed increasing, but on the other hand it can hardly be supposed that the effects on demand resulting from the demonetisation of silver have already been exhausted. The Indian Mint was closed and the Sherman Act repealed only last year, and trade is still suffering.

How long will the available supply of gold remain sufficient for our wants after business revives?—I am, Sir, &c.,

X.

London, Sept. 12th.

LIBRARY OF THE UNIVERSITY OF CALIFORNIA

www.ingramcontent.com/pod-product-compliance
Lightning Source LLC
Chambersburg PA
CBHW021934190326
41519CB00009B/1019